Huiling Zhang

2008 Feb.

PRINCIPLES OF

TOXICOLOGY TESTING

PRINCIPLES OF
TOXICOLOGY
TESTING

FRANK A. BARILE
St. John's University
Queens, New York

CRC Press
Taylor & Francis Group
Boca Raton   London   New York

CRC Press is an imprint of the
Taylor & Francis Group, an informa business

CRC Press
Taylor & Francis Group
6000 Broken Sound Parkway NW, Suite 300
Boca Raton, FL 33487-2742

© 2008 by Taylor & Francis Group, LLC
CRC Press is an imprint of Taylor & Francis Group, an Informa business

No claim to original U.S. Government works
Printed in the United States of America on acid-free paper
10 9 8 7 6 5 4 3 2 1

International Standard Book Number-10: 0-8493-9025-7 (Hardcover)
International Standard Book Number-13: 978-0-8493-9025-8 (Hardcover)

Library of Congress Cataloging-in-Publication Data

Barile, Frank A.
    Principles of toxicology testing / Frank A. Barile.
       p. ; cm.
    Includes bibliographical references and index.
    ISBN-13: 978-0-8493-9025-8 (hardcover : alk. paper)
    ISBN-10: 0-8493-9025-7 (hardcover : alk. paper)
    1. Toxicity testing. I. Title.
    [DNLM: 1. Toxicity Tests--methods. 2. Toxicology--methods. 3. Pharmacology, Clinical--methods. 4. Toxicity Tests--standards. QV 602 B252p 2007]

RA1199.B35 2007
615.9'07--dc22
                                         2007000174

**Visit the Taylor & Francis Web site at
http://www.taylorandfrancis.com**

**and the CRC Press Web site at
http://www.crcpress.com**

Dedication

To Pauline

Contents

SECTION I Basic Concepts in Toxicology Testing

SECTION II   Toxicology Testing In Vivo

SECTION III Toxicology Testing In Vitro

Preface

In the past few decades, the science of toxicology testing has evolved from an applied and supportive science, progressing in the shadows of more highly defined areas of mechanistic toxicology, to its own refined and technical discipline. Much like the development of other fields such as genetic engineering or information technology, the maturing of the discipline has been sporadic. In particular, its progress has been prompted particularly by public health initiatives, needs, and responses. The emerging and continuous accessibility of chemicals that are ubiquitous in society and therapeutic drugs that have extended the human life span to unparalleled levels has prompted faster, more accurate, reliable, and economical methods for screening for potential toxicity. With the burgeoning advances of biotechnology comes the availability of corresponding *in vitro* systems that complement traditional animal toxicology testing methods.

This book is divided into three sections: Section I, Basic Concepts in Toxicology Testing; Section II, Toxicology Testing *In Vivo*; and Section III, Toxicology Testing *In Vitro*. It begins with an introduction into the fundamentals of toxicology (Section I) to prepare students for the subsequent topics, and continues with discussions of toxicokinetics and human risk assessment. This introductory material is useful in explaining the applications of toxicology testing.

Section II describes in detail the fundamental principles of toxicology testing in animals. It describes acute toxicity, subchronic, and chronic studies performed in animals. Special emphasis is placed on study design and determination of classical indicators of acute and chronic studies such as the LD_{50} values. Other short- and long-term animal toxicity testing methodologies including dermal, ocular, and reproductive toxicity testing are discussed. Mutagenicity and carcinogenicity studies are also discussed in separate chapters.

Section III introduces and discusses *in vitro* alternatives to animal toxicology tests. It emphasizes cell culture methodology, cellular methods for acute systemic toxicity, and target organ and local toxicity. The advantages and disadvantages of alternative methods are presented. Special features of this section describe the use of high throughput screening (HTS) and its applications, the concepts of standardization and validation of *in vitro* techniques (especially large organized validation efforts currently supported by United States and European Union regulatory agencies), and the theories supporting the development of *in vitro* methodologies. Undergraduate and graduate toxicology students and industrial and academic research laboratories will find the text useful for entry level students in the discipline or for establishing a toxicology testing laboratory, respectively.

The juxtaposition of the principles of animal toxicology testing in the same text as *in vitro* alternative methods highlights the importance of each field for interpretation of the significance and relevance of the other. Thus, the discussions continuously refer to the corresponding methods available and the potential results from complementary designs of studies. In fact, both animal and *in vitro* toxicology testing methods are currently employed, often together, in toxicological analysis, derivation of mechanisms of toxicity, mutagenicity testing, and preclinical drug development.

Several excellent texts are available concerning the details of individual protocols. Consequently, although some procedures are outlined, the emphasis here is on the principles of the disciplines rather than on the particular steps of the techniques. In fact, the title, *Principles of Toxicology* (rather than *Toxicity*) *Testing*, emphasizes the universal application of the field as a scientific discipline as opposed to elaboration of a laboratory technique. In addition, the book highlights contemporary issues in toxicology testing including the various means of possible exposure to chemicals, high throughput screening of chemicals for toxicity testing and preclinical drug development, and an overview of applications. Overall, the reader is challenged to interpret the significance of toxicology testing results and to construct a logical approach toward the ultimate purpose of testing. Thus, the information contained herein is presented with great enthusiasm, particularly for the students prepared to dedicate their careers to this intriguing and fascinating scientific discipline.

Acknowledgments

My appreciation is extended toward my colleagues, particularly Dr. Charles Ashby, Jr. and Dr. Raymond Ochs for their insights, and my graduate students, Sushil Pai and Anthony Calabro, for their efforts and perseverance. I am also thankful to the editorial staff at Taylor & Francis, Ltd., for their interest, professionalism, and commitment to the project.

Author

Frank A. Barile, Ph.D., is associate professor in the Toxicology Division of the Department of Pharmaceutical Sciences at St. John's University College of Pharmacy and Allied Health Professions, New York. He earned a B.S. in pharmacy (1977), an M.S. in pharmacology (1980), and a Ph.D. in toxicology (1982) from St. John's University. After completing a post-doctoral fellowship program in pulmonary pediatrics at the Albert Einstein College of Medicine, Bronx, New York, he moved to the Department of Pathology at St. Luke's Roosevelt Hospital of Columbia University as a research associate investigating the role of pulmonary toxicants on collagen metabolism in cultured lung cells. In 1984, he was appointed assistant professor in the Department of Natural Sciences at York College of the City University of New York. He rejoined St. John's University in the Department of Pharmaceutical Sciences in 2000 and became an instrumental part of the toxicology program in the College of Pharmacy.

Dr. Barile holds memberships in several professional associations including the United States Society of Toxicology, the American Association of University Professors, the American Association for the Advancement of Science, the American Society of Hospital Pharmacists, the New York City Pharmacists Society, the New York Academy of Sciences, and the New York State Council of Health System Pharmacists. He has been appointed a consultant scientist by several professional groups including Schneider's Children's Hospital, Long Island Jewish–Cornell Medical Centers, is a committee member of SACATM, ICCVAM/NICEATM, and NIEHS, and an elected officer of the *In Vitro* specialty section of the United States Society of Toxicology.

Dr. Barile has been the recipient of Public Health Service research grants from the National Institutes of Health (NIGMS) for the past 20 years and received awards from the Minority Biomedical Research Support (MBRS) Program, the Minority High School Student Research Apprentice Program (MHSSRAP), and the AREA Program.

Dr. Barile has authored and co-authored approximately 75 papers and abstracts in peer-reviewed biomedical and toxicology journals as well as two books and one contributed chapter. He contributed original *in vitro* toxicology data to the international Multicenter Evaluation for Cytotoxicity (MEIC) program. He lectures regularly to toxicology and pharmacy undergraduate and graduate students (he was named Professor of the Year for the College of Pharmacy by the University Student Government Association in 2003). Dr. Barile continues to perform fundamental research on the cytotoxic effects of environmental chemicals and therapeutic drugs on cultured human and mammalian cells.

Section I

**Basic Concepts in
Toxicology Testing**

1 Introduction to Principles of Toxicology

1.1 INTRODUCTION

From its inception in the medieval era to its maturity as a distinct and separate discipline in the 21st century, toxicology was taught primarily as an applied science. The field incorporated the various approaches from a variety of disciplines, not limited to the broad sciences of chemistry and biology. In particular, toxicology evolved from applications of analytical and clinical chemists whose job definitions included chemical identification and analysis of body fluids.

The first modern toxicologists were chemists who had specialized training in inorganic separation methods including chromatographic techniques. Analytical chemists later employed thin layer and gas chromatography. As the instrumentation evolved and the technology became more exacting, high performance liquid chromatography was incorporated into the analytical arsenal in order to isolate minute quantities of compounds from complex mixtures of toxicological importance. Eventually, biological applications exerted their influence, incorporating such specialties as microbiology, genetics, and cell culture methodology. Today, the field has burgeoned into areas of specialization, some of which are hardly identifiable with their ancestral origins.

1.2 TYPES OF TOXICOLOGY

1.2.1 GENERAL TOXICOLOGY

General toxicology involves studies of exposure to chemical, biological, or physical agents and their untoward consequences that affect biological systems. The term, however, has been replaced by descriptions of areas that more closely reflect the specialized fields of study within the discipline. The development of advanced methodologies in biotechnology, the requirement for increased training, and the involvement of toxicology in legal applications have made it necessary to more accurately label the discipline according to the expanding body of specialties. As a result, a variety of descriptions further define the field of toxicology.

1.2.2 MECHANISTIC TOXICOLOGY

Mechanistic toxicology involves the identification of the cause of toxicity of a chemical at the level of a cell, tissue, or organ. The classification of toxicity of a chemical, therefore, may be expressed in terms of its mechanism of toxicity, its site of action, target area, or the organ most affected by the toxic insult. A similar expression, *mechanism of action*, is universally applied in the study of pharmacology. Thus, mechanistic toxicology seeks to determine the biochemical, physiological, or organic basis of a toxic agent's effects on biological systems.

1.2.3 REGULATORY TOXICOLOGY

Regulatory toxicology relates to the administrative dogma associated with the potential exposure to toxic agents encountered in the environment, in occupational settings, and in the home. Regulatory toxicology defines, directs, and dictates the rate at which an individual may encounter a synthetic or naturally occurring toxin, and establishes guidelines for its maintenance in the environment or within the therapeutic market. The guidelines are generally promulgated by agencies whose jurisdiction and regulations are established by federal, state, and local authorities.

1.2.4 DESCRIPTIVE TOXICOLOGY AND CLINICAL TOXICOLOGY

Descriptive toxicology is a subjective attempt to explain toxic agents and their applications. The descriptive area developed principally as a method for bridging the vacuum between science and the public's understanding of the field, especially when it became necessary for non-scientific sectors to comprehend the importance of toxicology. For instance, the study of metals in the environment (metals toxicology) has become a popular discipline for toxicologists interested in examining the roles of heavy or trace metals in the environment. Clinical toxicology may also be considered a descriptive category, although toxic agents with clinical applications are also characterized in other toxicological fields. Table 1.1 defines other descriptive

TABLE 1.1
Other Descriptive Fields of Toxicology

Descriptive Field	Definition
Genetic toxicology	Incorporates molecular biology principles in applications of forensic sciences such as DNA testing applied as legal evidence in court proceedings or related to toxic agents that interfere with normal gene function
Occupational toxicology	Examines hazards associated with toxic exposures in workplaces
In vitro toxicology	Development of cell culture techniques as alternatives to animal toxicity testing; a more current definition is *alternative methods to animal toxicology,* that more accurately describes applications of *in vitro* methods
Analytical toxicology	Chemical and biochemical procedures and methods associated with identification, analysis, and detection of toxic substances in specimens
Developmental toxicology	Study of toxic substances and their potential effects on biological reproduction and embryogenesis
Immunotoxicology	Study of toxic substances and their potential effects on immunity and resistance
Neurotoxicology	Study of toxic substances and their potential effects on functions and activities of nervous systems

fields of toxicology. More recently, the field has blossomed into further broad descriptive areas including the study of apoptosis, receptor-mediated signal transduction, gene expression, proteomics, oxidative stress, and toxicogenomics, among others.

1.3 COMMON TERMS AND NOMENCLATURE

The classic definition of toxicology has traditionally been understood as the study of xenobiotics, the science of poisons, and refers particularly to the interactions of exogenous agents with biological systems. For purposes of organizing the nomenclature, chemicals, compounds, and drugs are often referred to as *agents* or *substances*. Because such agents induce undesirable effects, they are usually alluded to as *toxins*. Consequently, toxicology involves internal and external physiological exposures to toxins and the interactions of toxins with organisms. Gradually, the term has evolved to include many chemically or physically unrelated classes of agents. What transforms a chemical into a toxin, therefore, depends more on the duration of exposure, dose (or concentration), or route of exposure, and less on the chemical structure, product formulation, or intended use of the material. As a result, almost any agent has the potential for toxicity, and thus falls within the broad definition of toxicology.

1.4 APPLICATIONS OF TOXICOLOGY

1.4.1 RESEARCH

1.4.1.1 Academic Applications

In the academic arena, research toxicologists examine the broad issues of toxicology in the laboratory setting. Academic concerns include all the public health areas in which progress in understanding toxicological sciences is necessary and include the elucidation of mechanistic, clinical, and descriptive toxicological theories. Research methods are modified to answer specific questions that arise from toxicological concerns that affect public health.

1.4.1.2 Industrial Applications

Research toxicologists employed in toxicity testing in the biotechnology and pharmaceutical industries perform toxicity testing — the screening of chemicals and drugs that have toxic potential before they are marketed. Preclinical testing in the pharmaceutical industry involves the conduct of Phase I trials to test the toxicities of candidate agents that have been chemically and biochemically screened as potentially useful therapeutic drugs. The toxicity testing procedures include both *in vitro* and animal protocols.

1.4.2 REGULATORY TOXICOLOGY

Regulatory toxicologists are employed primarily in government administrative agencies, as consultants to government and industry, or as representatives of industrial concerns. In this role, they sanction, approve, and monitor the uses of chemicals by establishing rules and guidelines. The guiding principles are promulgated through laws enacted by appropriate federal, state, and local jurisdictions that grant regulatory agencies their authority. Thus, through these regulations, an agency determines who is accountable and responsible for manufacturing, procurement, distribution, marketing, and ultimately, release and dispensing of chemical substances to the public.

1.4.3 FORENSIC TOXICOLOGY

Forensic toxicologists integrate appropriate techniques to identify compounds arising from mixtures of sometimes unrelated poisons as a result of incidental or deliberate exposure. Initially, forensic sciences profited from the application of the principles of chemical separation methods for the identification of controlled substances in body fluids. Later, forensic toxicologists applied biological principles of antigen–antibody interaction for paternity testing. By using the principles of blood grouping and the exclusion of the possible outcomes of paternal contributions to offspring phenotypes, it became possible to eliminate a male as a possible father of a child.

Antigen–antibody interactions also became the bases for enzyme-linked immunosorbent assays (ELISA, EMIT), currently used for specific and sensitive identification of drugs in biological fluids. Radioimmunoassays (RIAs) utilize similar antigen–antibody reactions while incorporating radiolabeled ligands as indicators. DNA separation and sequencing techniques have now almost totally replaced traditional paternity exclusion testing. These methods are also the basis for inclusion or exclusion of evidence in criminal and civil cases.

1.4.4 CLINICAL TOXICOLOGY

Clinical toxicologists have evolved and branched away from their forensic counterparts. The clinical toxicologist is interested in identification, diagnosis, and treatment of a condition, pathology, or disease resulting from environmental, therapeutic, or illicit exposure to chemicals or drugs. Exposure is commonly understood to include individual risk of contact with a toxin but can further be defined to include population risk.

1.5 CLASSIFICATION OF TOXIC AGENTS

Classification of toxic agents is a daunting task, considering the vast numbers and complexities of chemical compounds in the public domain. The availability of the variety of chemicals, drugs, and physical agents, along with their varied toxicological and pharmacological effects, means that a single agent may be listed in several different categories. Even compounds with similar structures or toxicological actions may be alternatively grouped according to their activities or physical states. The following outline of the classification system is generally accepted for demonstrating the complex nature of toxins.

TABLE 1.2
Classification of Pesticides

Pesticide Class	Classification According to Target
Insecticide	Organophosphorus esters
	Organochlorine compounds
	Carbamate esters
	Pyrethroid esters
	Botanical derivatives
Herbicide	Chlorphenoxy compounds
	Bipyridyl derivatives
	Chloroacetanilides
	Phosphonomethyl amino acids
Rodenticide	Anticoagulants
	α-Naphthyl thiourea
	Miscellaneous metals, inorganics, natural products
Fungicide	Agricultural, domestic, and therapeutic antifungals

1.5.1 CLASSIFICATION ACCORDING TO USE

1.5.1.1 Pesticides

The United States Environmental Protection Agency (EPA) defines a pesticide as a substance or mixture of substances intended to prevent, destroy, repel, or mitigate a pest. In general, pesticides are classified according to their biological targets. The four major classes of pesticides are insecticides, herbicides, rodenticides, and fungicides. Table 1.2 lists these categories and their general subclasses. Because of the physiological and biochemical similarities of target species and mammalian organisms, an inherent toxicity is associated with pesticides in mammalian organisms. In addition, within each classification, compounds are identified according to mechanism of action, chemical structure, or semi-synthetic source. For instance, although many fungicide categories exist, fungicidal toxicity in humans is mostly low order, with the exception of therapeutic anti-fungals, principally because of their specific mechanisms of action. Similarly, fumigants range from carbon tetrachloride to ethylene oxide and are used to kill insects, roundworms, and fungi in soil, stored grain, fruits, and vegetables. Their toxicity, however, is limited to occasional occupational exposure.

1.5.1.2 Food and Industrial Additives

Direct food and color additives are intentionally incorporated in foods and food processing operations for purposes of changing, enhancing, or masking color. They are also used for a variety of functionalities ranging from anti-caking agents to stabilizers, thickeners, and texturizers. Food and industrial additives fall into the field of food toxicology and readers are referred to review articles listed at the end of this chapter for information concerning food ingredients and contaminants.

1.5.1.3 Therapeutic Drugs

Toxicological classification of therapeutic agents follows their pharmacological mechanisms of action or their principal target organs of toxicity. Several important references address clinical toxicologies of therapeutic drugs as extensions of their adverse reactions and direct effects resulting from their excessive use.

1.6 SOURCES OF TOXINS

1.6.1 BOTANICAL

Contact dermatitis caused by poison ivy is a well characterized syndrome of acute inflammation. Today, many compounds of botanical origin are classified as herbal supplements, implying that their origins are botanical. Their importance in maintaining health is also related to their natural derivations. The toxicities of these agents, however, are poorly understood.

1.6.2 ENVIRONMENTAL

As a result of industrialization, many chemicals are associated with and classified according to their continuous presence in the environment, that is, water, land, and soil. The phenomenon is not limited to Western developed nations and is becoming a problem among developing South American, Asian, and African countries.

Environmental toxicology is a distinct discipline encompassing the areas of air pollution and ecotoxicology. A discussion of air pollution necessarily includes outdoor and indoor air pollution, presence of atmospheric sulfuric acid, airborne particulate matter, interaction of photochemicals with the environment, and chemicals found in smog. *Ecotoxicology* is the branch of environmental toxicology that investigates the effects of environmental chemicals on the ecosystems in question. Readers are referred to the review articles at the end of this chapter for further discussion.

SUGGESTED READINGS

Abraham, J., The science and politics of medicines control, *Drug Saf.*, 26, 135, 2003.
Bois, F.Y., Applications of population approaches in toxicology, *Toxicol. Lett.*, 120, 385, 2001.
Eaton, D.L. and Klassen, C.D., Principles of toxicology, in *Casarett and Doull's Toxicology: The Basic Science of Poisons*, Klassen, C.D., Ed., 6th ed., McGraw-Hill, New York, 2001, ch. 2.
Ettlin, R.A. et al., Careers in toxicology in Europe: options and requirements. Report of a workshop presented at the EUROTOX Congress in London (September 17–20, 2000), *Arch. Toxicol.*, 75, 251, 2001.
Gennings, C., On testing for drug/chemical interactions: definitions and inference, *J. Biopharm. Stat.*, 10, 457, 2000.
Greenberg, G., Internet resources for occupational and environmental health professionals, *Toxicology*, 178, 263, 2002.

Guzelian, P.S., Victoroff, M.S., Halmes, N.C., James, R.C., and Guzelian, C.P., Evidence-based toxicology: a comprehensive framework for causation, *Hum. Exp. Toxicol.* 24, 161, 2005.

Kaiser, J., Toxicology: tying genetics to the risk of environmental diseases, *Science,* 300, 563, 2003.

Meyer, O., Testing and assessment strategies including alternative and new approaches, *Toxicol. Lett.,* 140, 2003.

Tennant, R.W., The National Center for Toxicogenomics: using new technologies to inform mechanistic toxicology, *Environ. Health Perspect.,* 110, A8, 2002.

REVIEW ARTICLES

Aardema, M.J. and MacGregor, J.T., Toxicology and genetic toxicology in the new era of "toxicogenomics": impact of "-omics" technologies, *Mutat. Res.,* 499, 13, 2002.

Abraham, J. and Reed, T., Progress, innovation and regulatory science in drug development: the politics of international standard setting, *Soc. Stud. Sci.,* 32, 337, 2002.

Barnard, R.C., Some regulatory definitions of risk: interaction of scientific and legal principles, *Regul. Toxicol. Pharmacol.,* 11, 201, 1990.

Collins, T.F., History and evolution of reproductive and developmental toxicology guidelines, *Curr. Pharm. Des.* 12, 1449, 2006.

Eason, C. and O'Halloran, K., Biomarkers in toxicology versus ecological risk assessment, *Toxicology,* 181, 517, 2002.

Fostel, J. et al., Chemical effects in biological systems data dictionary (CEBS-DD): a compendium of terms for the capture and integration of biological study design description, conventional phenotypes, and "omics" data. *Toxicol. Sci.* 88, 585, 2005.

Harris, S.B. and Fan, A.M., Hot topics in toxicology, *Int. J. Toxicol.,* 21, 383, 2002.

Johnson, D.E. and Wolfgang, G.H., Assessing the potential toxicity of new pharmaceuticals, *Curr. Top. Med. Chem.,* 1, 233, 2001.

Lewis, R.W. et al., Recognition of adverse and nonadverse effects in toxicity studies, *Toxicol. Pathol.,* 30, 66, 2002.

Reynolds, V.L., Applications of emerging technologies in toxicology and safety assessment, *Int. J. Toxicol.,* 24, 135, 2005.

Rietjens, I.M. and Alink, G.M., Future of toxicology: low dose toxicology and risk–benefit analysis, *Chem Res Toxicol.* 19, 977, 2006.

Schrenk, D., Regulatory toxicology: objectives and tasks defined by the working group of the German society of experimental and clinical pharmacology and toxicology, *Toxicol. Lett.,* 126, 167, 2002.

Waddell, W.J., The science of toxicology and its relevance to MCS, *Regul. Toxicol. Pharmacol.,* 18, 13, 1993.

Wolfgang, G.H. and Johnson, D.E., Web resources for drug toxicity, *Toxicology,* 173, 67, 2002.

2 Effects of Chemicals

2.1 TOXICOLOGICAL EFFECTS

2.1.1 GENERAL CLASSIFICATION

As noted in Chapter 1, chemical and physical agents are categorized in part according to several indicators of toxicity that render them amenable for understanding their potential toxic effects. Prediction of toxicity of a chemical lies in the ability to recognize its potential adverse effects based on factors not necessarily related to physicochemical properties. This chapter explores a variety of local and systemic reactions elicited by chemical exposure and the physiological and immunological basis of such effects.

2.1.2 CHEMICAL ALLERGIES

The four types of immunological hypersensitivity (allergic) reactions include:

1. Type I Antibody-mediated reactions
2. Type II Antibody-mediated cytotoxic reactions
3. Type III Immune complex reactions
4. Type IV Delayed-type hypersensitivity cell-mediated immunity

Type I antibody-mediated reactions occur in three phases. The initial or sensitization phase is triggered by contact with a previously unrecognized antigen. This reaction entails binding of the antigen to the immunoglobulin E (IgE) present on the surfaces of mast cells and basophils. The second or activation phase follows after an additional dermal or mucosal challenge with the same antigen. This phase is characterized by degranulation of mast cells and basophils, with a subsequent release of histamine and other soluble mediators. The third stage, the effector phase, is characterized by accumulation of pre-formed and newly synthesized chemical mediators that precipitate local and systemic effects. Degranulation of neutrophils and eosinophils completes the late-phase cellular response.

Antigens involved in type I reactions are generally airborne pollens including mold spores and ragweed as well as food ingredients. Ambient factors such as heat and cold, drugs (opioids, antibiotics) and metals (silver, gold) precipitate chemical allergies of the type I nature. Because of their small molecular weights, the majority of chemicals and drugs, as single entities, generally circulate undetected by immune surveillance systems. Consequently, in order to initiate the sensitization phase of an

TABLE 2.1
Examples of Type I Hypersensitivity Syndromes, Causes, and Effects

Syndrome	Causes	Effects
Allergic rhinitis (hay fever)	Pollens, mold spores	↑ capillary permeability in nasal and frontal sinuses and musosal membranes, ↑ vasodilation
Food allergies	Lectins and proteins in nuts, eggs, shellfish, dairy products	↑ capillary permeability, ↑ vasodilation, ↑ smooth muscle contraction
Atopic (allergic) dermatitis	Localized exposure to drugs and chemicals	Initial local mast cell release of cytokines followed by activation of neutrophils and eosinophils
Asthma	Chemicals, environmental, behavioral factors	Chronic obstructive reaction of LRT involving airway hyperactivity and cytokine release

LRT = lower respiratory tract.

antigenic response, chemicals are immunologically handled as haptens.* Some examples of type I hypersensitivity syndromes are described in Table 2.1 and typical effects of chemical allergies are listed in Table 2.2. The effects of chemical allergies are usually acute and appear immediately when compared to other hypersensitivity reactions described below.

Type II antibody-mediated cytotoxic reactions differ from type I in the nature of antigen, the cytotoxic character of the antigen–antibody reaction, and the type of antibody formed (IgM, IgG). In general, antibodies are induced against target antigens that are altered cell membrane determinants. Type II reactions include complement–mediated (CM) reactions, antibody-dependent cell-mediated cytotoxicity (ADCC), antibody-mediated cellular dysfunction (AMCD), transfusion reactions, Rh incompatibility reactions, autoimmune reactions, and chemical-induced reactions, the last of which is of greater interest in experimental toxicology.

As with type I reactions, chemical-induced type II cytotoxicity requires that an agent behave as a hapten. The chemical binds to the target cell membrane and proceeds to operate as an altered cell membrane determinant. This determinant changes the conformational appearance of a component of the cell surface, not unlike the effect described above for a hapten. Thus, it induces a series of responses that terminate in antibody induction. The determinant attracts a variety of immune surveillance reactions including a complement-mediated cytotoxic reaction, recruitment of granulocytes, or deposit of immune complexes within the cell membrane. A variety of therapeutic drugs are known to induce type II reactions, particularly with continuous administration. Examples include antibiotics and cardiovascular agents, among others.

* Haptens are small molecular weight chemical entities (<1000 D) that non-specifically bind to larger circulating polypeptides or glycoproteins. Binding induces a conformational change in the original larger complex. The chemical entity bound to the larger molecule is no longer recognized as part of the host (self) system, rendering it susceptible to immune attack.

TABLE 2.2
Inflammatory Effects of Chemical Allergies

Effect	Description
Bullous	Large dermal blisters
Erythema	Redness or inflammation of skin due to dilation and congestion of superficial capillaries (e.g., sunburn)
Flare	Reddish, diffuse blushing of skin
Hyperemia	Increased blood flow to tissue or organ
Induration	Raised, hardened, thickened skin lesion
Macule	Flat red spot on skin due to increased blood flow
Papule	Raised red spot on skin due to increased blood flow and antibody localization
Petechial	Small, pinpoint dermal hemorrhagic spots
Pruritis	Itching
Purulent	Suppuration; production of pus containing necrotic tissue, bacteria, inflammatory cells
Urticaria	Pruritic skin eruptions characterized by wheal formation
Vesicular	Small dermal blisters
Wheal	Raised skin lesion due to accumulation of interstitial fluid

Type III immune complex reactions are localized responses mediated by antigen–antibody immune complexes. Type III reactions are stimulated by microorganisms and involve activation of complement. Systemic (serum sickness) and localized (arthus reaction) immune complex disease, infection-associated immune complex disease (rheumatic fever), and occupational diseases (opportunistic pulmonary fungal infections) induce complement–antibody–antigen complexes that trigger the release of cytokines and recruitment of granulocytes, resulting in increased vascular permeability and tissue necrosis.

Type IV (delayed-type) hypersensitivity cell-mediated immunity involves antigen-specific T cell activation. The reaction starts with an intradermal or mucosal challenge (sensitization stage). CD4+ T cells then recognize MHC-II (major histocompatibility class-II) antigens on antigen-presenting cells (such as Langerhans cells) and differentiate to T_H1 cells. This sensitization stage requires prolonged local contact with the agent, from several days to weeks. A subsequent repeat challenge stage induces differentiated T_H1 (memory) cells to release cytokines, further stimulating attraction of phagocytic monocytes and granulocytes. The release of lysosomal enzymes from the phagocytes results in local tissue necrosis. Contact hypersensitivity resulting from prolonged exposure to natural products and metals, for example, is caused by the lipophilicity of the chemicals in skin secretions, thus acting as a hapten.

2.1.3 IDIOSYNCRATIC REACTIONS

Idiosyncratic reactions are abnormal responses to chemicals or drugs generally resulting from uncommon genetic predisposition. An exaggerated response to the skeletal muscle relaxant properties of succinylcholine, a depolarizing neuromuscular blocker, classifies as a typical idiosyncratic reaction. In some subjects, a congenital deficiency in plasma cholinesterase results in a reduction in the rate of succinylcholine deactivation. As succinylcholine accumulates, respiration fails to return to normal during the postoperative period. Similarly, the cardiotoxic action of cocaine is exaggerated in cases of congenital deficiency of plasma esterases, which are necessary for metabolism of the drug. A paucity of circulating enzymes allows for uncontrolled, sympathetically mediated effects in experimental animals and humans.

2.1.4 IMMEDIATE VERSUS DELAYED HYPERSENSITIVITY

In contrast to immune hypersensitivity reactions, some chemical effects are immediate or delayed, depending on the mechanism of toxicity. The acute effects of sedative hypnotics are of immediate consequence — high concentrations in animals or humans result in death from respiratory depression. The effects of carcinogens, however, may not be demonstrated for generations in humans or only after several years of exposure in rodents. An important example of this has been demonstrated with the link between diethylstilbesterol (DES) administration to child-bearing women in the 1950s and the subsequent development of clear cell adenocarcinoma vaginal cancers in the offspring.

2.1.5 REVERSIBLE VERSUS IRREVERSIBLE EFFECTS

In general, the effects of most chemicals or drugs are reversible until a critical point is reached — that is, when vital function is compromised or a mutagenic, teratogenic, or carcinogenic effect develops. In fact, the carcinogenic effects of chemicals such as those present in tobacco smoke may be delayed for decades until irreversible cellular transformation occurs. Reversibility of the acute effects of chemicals is achieved through the administration of antagonists, by enhancement of metabolism or elimination, by delaying absorption, by intervening with another toxicological procedure that decreases toxic blood concentrations, or by terminating the exposure.

2.1.6 LOCAL VERSUS SYSTEMIC EFFECTS

As discussed below, local or systemic effects of a compound depend on site of exposure. The integument (skin) and lungs are targets of chemical exposure because they frequently serve as the first sites of contact with environmental chemicals. Oral exposure requires absorption and distribution of an agent prior to the development of systemic effects. Hypersensitivity reactions types I and IV are precipitated by local activation of immune responses following a sensitization phase, while chemical-induced type II reactions are elicited through oral or parenteral administration.

2.1.7 MUTAGENIC AND CARCINOGENIC EFFECTS

Mutagenic and carcinogenic effects of chemical exposure are discussed in detail in Part II, Chapter 10.

2.2 BIOCHEMICAL PROPERTIES

2.2.1 CHEMICAL STRUCTURE

Important components of a toxicity testing protocol include the identification, categorization, synthesis, and toxicity screening of chemicals according to their classifications. Because chemical and drug development depends upon the availability of parent compounds, the synthesis of structurally related compounds is an important process. Thus, understanding the chemical structure of a compound allows for reasonable prediction of many of its anticipated and adverse effects. Some examples of agents that are categorized according to their molecular structures include organophosphorus insecticides, heavy metals, benzodiazepines (sedative hypnotics), and imidazolines (tranquilizers).

2.2.2 MECHANISM OF ACTION

Similarly, it is convenient to organize toxic agents according to their physiologic or biochemical targets. Examples include mutagens, hepatotoxic compounds, methemoglobin-producing agents, and acetylcholinesterase inhibitors.

2.3 EXPOSURE

In the presence of specific circumstances, any chemical has the potential for toxicity — i.e., the same dose of a chemical may be harmless if limited to oral exposure but toxic if inhaled or administered parenterally. Thus the route and site of exposure exert significant influence in determining the toxicity of a substance. More frequently, a therapeutic dose for an adult may be toxic for an infant or child. Similarly, a substance may not exert adverse effects until a critical threshold is achieved. Thus, in order to induce toxicity, it is necessary for the chemical to accumulate in a physiological compartment at a concentration sufficiently high to reach the threshold value. Finally, repeated administration over a specified period of time also determines the potential for toxicity. The following discussion details the circumstances for exposure and dose of a chemical that favors or deters the potential for toxicity.

2.3.1 ROUTE

2.3.1.1 Oral Administration

By far, oral administration of toxins is the most popular route of exposure. Oral administration involves the presence of several physiological barriers that must be penetrated or circumvented if adequate blood concentrations of a compound is achieved (Figure 2.1). The mucosal layers of the oral cavity, pharynx, and esophagus consist of stratified squamous epithelium that serves to protect the upper gastrointestinal (GI) lining from the effects of contact with physical and chemical agents. Simple columnar epithelium lines the stomach and villi of the intestinal tract that function in digestion, secretion, and absorption. Immediately underlying the epithelium is the lamina propria, a mucosal layer rich in blood vessels and nerves. Mucosa-

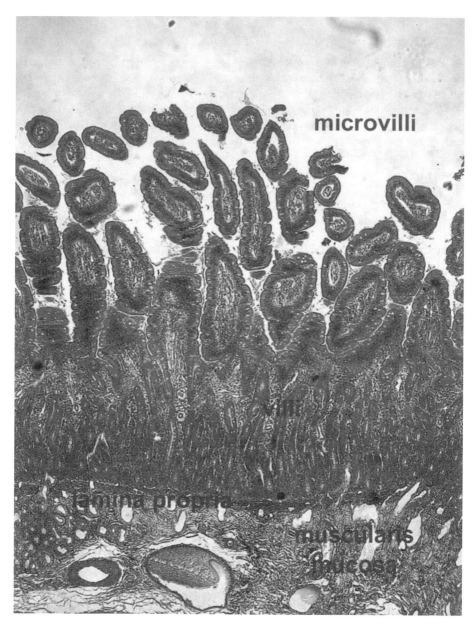

FIGURE 2.1 Histological slide of intestinal villi in longitudinal section (villi) and cross section (rounded structures above). The anatomy of the small intestine illustrates the interface of lumincal contents with the microvilli, composed of simple columnar epithelial cells that coat the surface of the villi. The microvillus membrane increases the surface area of the intestinal tract 600-fold to allow for sufficient secretion and absorption. (Photo courtesy of Dr. Diane Hardej.)

associated lymphoid tissue (MALT) is layered within this level, where prominent lymphatic nodules sustain the presence of phagocytic macrophages and granulocytes. Salivary and intestinal glands contribute to the digestive process by secreting saliva and digestive juices. The submucosa, muscularis, and serosa complete the strata that form the anatomical envelope of the GI tract. Enteroendocrine and exocrine cells in the GI tract secrete hormones and, in the stomach, secrete acid and gastric lipase.

The primary function of the stomach is mechanical and chemical digestion of food, while absorption is secondary. Several factors influence the transit and stability of a chemical in the stomach, thereby influencing gastric emptying time (GET). The presence of food delays absorption and dilutes the contents of the stomach, thus reducing subsequent chemical transit. An increase in the relative pH of the stomach causes a negative feedback inhibition of stomach churning and motility, which also results in delay of gastric emptying. Any factor that slows stomach motility will increase the amount of time in the stomach, prolonging the GET. Thus, the longer the GET, the greater the duration of a chemical presence within the stomach, and the more susceptibility to gastric enzyme degradation and acid hydrolysis. In addition, prolonged GET delays passage to and subsequent absorption in the intestinal tract.

2.3.1.2 Intranasal Administration

Intranasal insufflation is a popular method for therapeutic administration of corticosteroids and sympathomimetic amines and for the illicit use of drugs of abuse such as cocaine and opioids. The dosage form in therapeutic use is usually aerosolized in a metered nasal inhaler. In the case of illicit use, crude drugs are inhaled (snorted) through the nares as fine or coarse powders. In both cases, absorption is rapid due to the extensive network of capillaries in the lamina propria of the mucosal lining within the nasopharynx. Thus the absorption rate rivals that of pulmonary inhalation.

2.3.1.3 Inhalation

The vast surface area of the upper and lower respiratory tracts allows for wide and immediate distribution of inhaled powders, particulates, aerosols, and gases. Figure 2.2 illustrates the thin alveolar wall that separates airborne particulates from access to capillary membranes. Once a drug is ventilated to the alveoli, it is transported across the alveolar epithelial lining to the capillaries, resulting in rapid absorption.

2.3.1.4 Dermal and Parenteral Routes

Parenteral administration includes epidermal, intradermal, transdermal, subcutaneous, intramuscular, and intravenous injections. Parenteral routes, in general, are subject to minimal initial enzymatic degradation or chemical neutralization, thus bypassing the hindrances associated with passage through epithelial barriers. Figure 2.3 illustrates the layers of the skin: the epidermis, the outermost layer, and the underlying dermis and subcutaneous (hypodermis) layers. Epidermal and upper

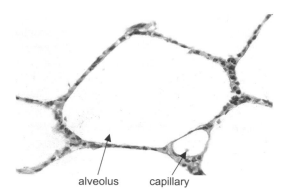

FIGURE 2.2 Pulmonary alveolus showing thin capillary–air interface. (From Barile, Frank A., *Clinical Toxicology: Principles and Mechanisms,* CRC Press, 2004.)

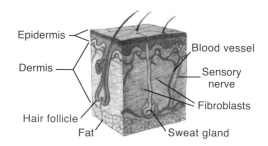

FIGURE 2.3 Cross-section of dermal epithelium and underlying basal lamina structures. Epidermis, dermis and underlying (subcutaneous) layers of skin. (From Artificial Skin, News Features and Articles, NIH, NIGMS, 2003.)

dermal injections have the poorest absorption capabilities of the parenteral routes primarily because of limited circulation.

The deeper dermal and subcutaneous layers provide entrance to a richer supply of venules and arterioles. A subcutaneous injection forms a depot within the residing adipose tissue with subsequent leakage into the systemic circulation or the majority of the injection can enter the arterioles and venules. Intramuscular injections ensure access to a more extensive vascular network within skeletal muscle, accounting for more rapid exposure than subcutaneous. Intravenous injection is the most rapid method of chemical exposure because access to the circulation is direct and immediate. Thus, the route of exposure contributes as much to toxicity as the dose.

2.3.2 DURATION AND FREQUENCY

2.3.2.1 Acute Exposure

In general, any exposure shorter than or equal to 24-hr is regarded as acute. Exposure to most toxic gases (carbon monoxide, hydrogen cyanide) requires less than 24 hr for toxicity. In most experimental settings, however, 72 hr may still be considered acute exposure, such as in continuous low-dose exposure to hepatotoxic agents. In addition, a single intravenous injection of a chemical is certainly classified as an acute exposure. Subacute exposure generally refers to continuous or repeated exposure to a chemical for more than 72 hr but less than 1 mo.

2.3.2.2 Chronic Exposure

Chronic exposure is any relative time period for which continuous or repeated exposure beyond the acute phase is required for the same chemical to induce a toxic response. Subchronic exposure is understood to involve a duration between acute and chronic. The traditional time of subchronic exposure is accepted as 1 to 3 mo. In experimental and clinical toxicity, however, a subchronic exposure may include repeated exposure for a period longer than 3 mo. Thus, the terms are flexible adaptations to define the onset of chemical intoxication. In addition, considerable overlap in judgment is involved when labels are assigned to exposure periods.

Frequency of administration involves repeated doses of a drug or toxin during the exposure period. Repeated administration of the same dose of a chemical within a time period defined as acute or chronic establishes a greater potential for adverse effects. Similarly, continuous repeated exposure to a toxin, especially during an acute time period, has greater toxic potential.

2.3.3 ACCUMULATION

Dose, duration, frequency, and route of exposure contribute to chemical toxicity in part through accumulation of a compound in physiological compartments. A normal dosage schedule is determined according to a chemical's half-life ($t_{1/2}$) in plasma and its intended response — that is, the time required for plasma levels to decrease to one-half of the measured or estimated concentration. Thus, if the frequency of administration occurs often within the $t_{1/2}$ of a chemical, its concentration in a compartment is likely to increase beyond the desirable level. Accumulation results from overloading of a chemical within this compartment.

2.3.3.1 According to Physiological Compartment

Biological systems in their entirety are considered one-compartment models. Ideally, a chemical capable of distributing uniformly throughout the body would maintain steady-state levels for the exposure period. Blood levels* of a compound may also decrease uniformly, assuming a constant rate of elimination. The body, however, is not a homogeneous chamber. A chemical, once absorbed, can distribute and/or bind to one or more of many physiological sites. The distribution of a chemical depends largely on its physicochemical characteristics (see Chapter 3, Toxicokinetics, for detailed descriptions of absorption, distribution, metabolism, and elimination).

The compartments include whole blood, serum and serum proteins, plasma and plasma proteins, adipose tissue, interstitial and extracellular fluids, alveolar air space, and bone marrow. In addition, a chemical may preferentially accumulate in any tissue or organ, thus acting as a discrete compartment. For instance, many

---

* The *blood level* and *plasma level* terms are often used interchangeably, although blood and plasma are distinct anatomical compartments.

therapeutic drugs such as warfarin (a vitamin K antagonist anticoagulant) non-specifically bind to circulating plasma proteins, resulting in apparently lower blood concentrations than anticipated. Heavy metals preferentially accumulate in adipose tissue, kidney, and bone. Consequently their toxicity may be experienced for prolonged periods as the compounds are slowly released from this compartment years after exposure has ceased. Accumulation, therefore, is predicted based on a chemical's apparent volume of distribution (V_d), which is estimated as the total dose of chemical in the body divided by the concentration of chemical in the plasma for a given period. In general, the greater the V_d, the greater the potential for accumulation in some physiological compartment.*

2.3.3.2 According to Chemical Structure

Accumulation is also determined by a chemical's structure and its interaction within the physiological compartment. This phenomenon is guided by the chemical's predominant state of existence in a physiological fluid — that is, it remains in the fluid compartment as an ionic or non-ionic species. In general, at physiological pH, lipid-soluble compounds will preferentially remain in their non-ionic states, to bind to and accumulate in membranes of tissues and organs. Conversely, water-soluble compounds remain as ionic species at the pH of blood. Thus, because they are less prone to tissue binding, the ions are readily available for renal secretion and elimination (see Chapter 3, Toxicokinetics, for a complete discussion).

2.4 CHEMICAL INTERACTIONS

2.4.1 POTENTIATION

Potentiation of toxicity occurs when the toxic effect of one chemical is enhanced in the presence of a toxicologically unrelated agent. The situation can be described numerically as $0 + 2 > 2$, where a relatively non-toxic chemical alone has little or no effect (0) on a target organ, but may enhance the toxicity of another co-administered chemical (2). The hepatotoxicity of carbon tetrachloride, for instance, is greatly enhanced in the presence of isopropanol.

2.4.2 ADDITIVE EFFECTS

Two or more chemicals whose combined effects are equal to the sum of the individual effects is described as an additive interaction. This is the case with the additive effects of a combination of sedative hypnotics and ethanol (drowsiness, respiratory depression). Numerically this is summarized as $2 + 2 = 4$.

2.4.3 SYNERGISTIC EFFECTS

By definition, a synergistic effect is indistinguishable from potentiation except that, in some references, both chemicals must have some cytotoxic activity. Numer-

---

* A standard measure for the accumulated internal dose of a chemical is *body burden,* which refers to the amount of chemical stored in one or several physiological compartments or in the body as a whole.

ically, synergism occurs when the sum of the effects of two chemicals is greater than the additive effects, such as the effect experienced with a combination of ethanol and antihistamines $(1 + 2 > 3)$. The synergism and potentiation terms often are used synonymously.

2.4.4 ANTAGONISTIC EFFECTS

The opposing actions of two or more chemical agents, not necessarily administered simultaneously, are considered antagonistic interactions. Different types of antagonism include:

> **Functional antagonism** — The opposing physiological effects of chemicals, such as with central nervous system stimulants versus depressants.
> **Chemical antagonism** — Drugs or chemicals that bind to, inactivate, or neutralize target compounds such as the actions of chelators in metal poisoning.
> **Dispositional antagonism** — The interference of one agent with the absorption, distribution, metabolism, or excretion (ADME) of another; examples of agents that interfere with absorption, metabolism, and excretion include activated charcoal, phenobarbital, and diuretics, respectively.
> **Receptor antagonism** — The occupation of toxicological or pharmacological receptors by competitive or non-competitive agents, such as the use of tamoxifen in the prevention of estrogen-induced breast cancer.

2.5 DOSE–RESPONSE RELATIONSHIP

2.5.1 GENERAL ASSUMPTIONS

A discussion of the effects of a chemical as a result of exposure to a particular dose is necessarily followed by an explanation of the pathway by which that dose elicited the response. This relationship has traditionally been known as the dose–response relationship.* The result of exposure to the dose is any measurable, quantifiable, or observable indicator. The response depends on the quantity and route of chemical exposure or administration within a given period. Two types of dose–response relationships exist, depending on the numbers of subjects and doses tested.

2.5.1.1 Graded Dose–Response

The graded dose–response describes the relationship of an individual test subject or system to increasing and/or continuous doses of a chemical. Figure 2.4 illustrates the effects of increasing doses of several chemicals on cell proliferation *in vitro*. The concentration of the chemical is inversely proportional to the number of surviving cells in the cell culture system.

---

* In certain fields of toxicology, particularly mechanistic and *in vitro* systems, the term dose response is more accurately referred to as concentration effect. This change in terminology makes note of the specific effect on a measurable parameter that corresponds to a precise plasma concentration.

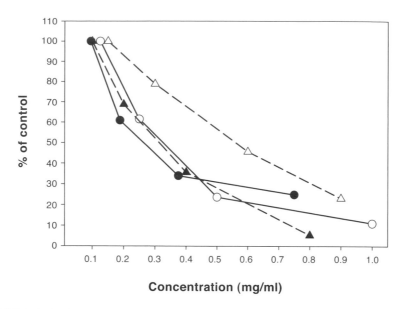

FIGURE 2.4 Graded dose–response curve for caffeine HCl (○), chloramphenicol HCl (●), atropine sulfate (△), and phenol (▲). Graph illustrates percent of viable human lung cells capable of proliferating in a cell culture system in response to increasing concentrations of chemicals. (From Yang A. et al., *Toxicol. in Vitro*, 16, 33, 2002.)

2.5.1.2 Quantal Dose–Response

Alternatively, the quantal dose–response is determined by the distribution of responses to increasing doses in a population of test subjects or systems. This relationship is generally classified as an "all-or-none effect" in which the test system or organisms are quantified as either responders or non-responders. A typical quantal dose–response curve is illustrated in Figure 2.5 by the LD_{50} (lethal dose 50%) distribution. The LD_{50} is a statistically calculated dose of a chemical that causes death in 50% of the animals tested. The doses administered are also continuous or at different levels and the response is generally mortality, gross injury, tumor formation, or some other criterion by which a standard deviation or cut-off value can be determined. In fact, other decisive factors such as therapeutic dose or toxic dose is determined using quantal dose–response curves, from which are derived the ED_{50} (effective dose 50%) and the TD_{50} (toxic dose 50%), respectively.

Graded and quantal curves are generated based on several assumptions. The time period at which the response is measured is chosen empirically or selected according to accepted toxicological practices. For instance, empirical time determinations may be established using a suspected toxic or lethal dose of a substance and the response is determined over several hours or days. This time period is then set for all determinations of the LD_{50} or TD_{50} for that period.* The frequency of administration is

---

* This time period is commonly set at 24 hr for LD_{50} determinations.

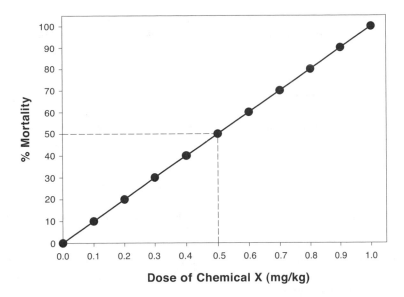

FIGURE 2.5 Quantal dose–response curve showing experimental derivation and graphic estimation of LD_{50}. (From Barile, Frank A., *Clinical Toxicology: Principles and Mechanisms,* CRC Press, 2004.)

assumed to be a single dose administered at the start of the time period when the test subjects are acclimated to the environment.

2.5.2 CONCENTRATION

It is assumed that at the conclusion of an experiment, the measured or observed effect is, in fact, due to the presence of the chemical. The establishment of this causal relationship is critical if valid conclusions are drawn from the dose–response curve. It is also presumed that the chemical in question is present at the receptor site or cellular target responsible for the effect. Support for this assumption follows from measurement of concentrations at the cellular or organ level. In fact, as the concentration of a chemical in the affected compartment increases, the degree of response must increase proportionately if this assumption is valid. For this reason, some references have suggested using the *concentration effect curve* as an alternative label to dose–response. The former is purported to be a more ingenuous description of the causal relationship and more accurately reflects the parameters measured.

2.5.3 CRITERIA FOR MEASUREMENT

Except for lethality, the selection of a measurable or observable endpoint is crucial. A desirable biomarker accurately reflects the presence of a chemical at a cellular or molecular site or suggests that a toxic effect originates from its action at the target organ. Selection of a measurable endpoint thus depends on the suspected mechanism of toxicity, if known, or on empirical determinations based on the chemical formula. Some biomarkers are also subjective, such as reliance on histological grading, calculation of the degree of anesthesia, pain, motor activity, or behavioral change. Thus the standards for quantifying the endpoint are determined and established prior to the experimental set-up.

SUGGESTED READINGS

Acquavella, J. et al., Epidemiologic studies of occupational pesticide exposure and cancer: regulatory risk assessments and biologic plausibility, *Ann. Epidemiol.*, 13, 1, 2003.

Ashby, J., The leading role and responsibility of the international scientific community in test development, *Toxicol. Lett.*, 140, 37, 2003.

Carnevali, O. and Maradonna, F., Exposure to xenobiotic compounds: looking for new biomarkers, *Gen. Comp. Endocrinol.*, 131, 203, 2003.

Eaton, D.L. and Klassen, C.D., Principles of toxicology, in *Casarett and Doull's Toxicology: The Basic Science of Poisons*, Klaassen, C.D., Ed., 6th ed., McGraw-Hill, New York, 2001, Ch. 2.

Olsen, C.M., Meussen-Elholm, E.T., Hongslo, J.K., Stenersen, J., and Tollefsen, K.E., Estrogenic effects of environmental chemicals: an interspecies comparison. *Comp. Biochem. Physiol. C Toxicol. Pharmacol.* 141, 267, 2005.

Yang, A., Cardona, D.L., and Barile, F.A., Subacute cytotoxicity testing with cultured human lung cells, *Toxicol. in Vitro,* 16, 33, 2002.

REVIEW ARTICLES

Basketter, D.A. et al., Factors affecting thresholds in allergic contact dermatitis: safety and regulatory considerations, *Contact Dermatitis*, 47, 1, 2002.

Bolt, H.M. and Kiesswetter, E., Is multiple chemical sensitivity a clinically defined entity? *Toxicol. Lett.*, 128, 99, 2002.

Dewhurst, I.C., Toxicological assessment of biological pesticides, *Toxicol. Lett.*, 120, 67, 2001.

Edler, L. and Kopp-Schneider, A., Statistical models for low dose exposure, *Mutat. Res.*, 405, 227, 1998.

Efroymson, R.A. and Murphy, D.L., Ecological risk assessment of multimedia hazardous air pollutants: estimating exposure and effects, *Sci. Total Environ.*, 274, 219, 2001.

Feron, V.J. et al., International issues on human health effects of exposure to chemical mixtures, *Environ. Health Perspect*, 110, 893, 2002.

Foster, J.R., The functions of cytokines and their uses in toxicology, *Int. J. Exp. Pathol.*, 82, 171, 2001.

Furtaw, E.J., Jr., An overview of human exposure modeling activities at the U.S. EPA's National Exposure Research Laboratory, *Toxicol. Ind. Health,* 17, 302, 2001.

Gaylor, D.W. and Kodell, R.L., Dose–response trend tests for tumorigenesis adjusted for differences in survival and body weight across doses, *Toxicol. Sci.*, 59, 219, 2001.

Goldman, L.R., Epidemiology in the regulatory arena, *M. J. Epidemiol.*, 154, 18, 2001.

Green, S., Goldberg, A., and Zurlo, J., TestSmart high production volume chemicals: an approach to implementing alternatives into regulatory toxicology, *Toxicol. Sci.*, 63, 6, 2001.

Hakkinen, P.J. and Green, D.K., Alternatives to animal testing: information resources via the Internet and World Wide Web, *Toxicology*, 173, 3, 2002.

Hanson, M.L. and Solomon, K.R., New technique for estimating thresholds of toxicity in ecological risk assessment, *Environ. Sci. Technol.*, 36, 3257, 2002.

Hastings, K.L., Implications of the new FDA/CDER immunotoxicology guidance for drugs, *Int. Immunopharmacol.*, 2, 1613, 2002.

Hattan, D.G. and Kahl, L.S., Current developments in food additive toxicology in the USA, *Toxicology*, 181, 417, 2002.

Herbarth, O. et al., Effect of sulfur dioxide and particulate pollutants on bronchitis in children: a risk analysis, *Environ. Toxicol.*, 16, 269, 2001.

Isbister, G.K., Data collection in clinical toxinology: debunking myths and developing diagnostic algorithms, *J. Toxicol. Clin. Toxicol.*, 40, 231, 2002.

Jarup, L., Health and environment information systems for exposure and disease mapping, and risk assessment. *Environ. Health Perspect.*, 112, 995, 2004.

Kenna, L.A., Labbe, L., Barrett, J.S., and Pfister, M., Modeling and simulation of adherence: approaches and applications in therapeutics, *AAPS J.*, 7, E390, 2005.

Kimber, I., Gerberick, G.F., and Basketter, D.A., Thresholds in contact sensitization: theoretical and practical considerations, *Food Chem. Toxicol.*, 37, 553, 1999.

Krishnan, K. and Johanson, G., Physiologically-based pharmacokinetic and toxicokinetic models in cancer risk assessment. *J. Environ. Sci. Health C Environ. Carcinog. Ecotoxicol. Rev.*, 23, 31, 2005.

Maurer, T., Skin as a target organ of immunotoxicity reactions, *Dev. Toxicol. Environ. Sci.*, 12, 147, 1986.

McCarty, L.S., Issues at the interface between ecology and toxicology, *Toxicology*, 181, 497, 2002.

McCurdy, T. et al., The national exposure research laboratory's consolidated human activity database, *J. Expo. Anal. Environ. Epidemiol.*, 10, 566, 2000.

Melnick, R.L. and Kohn, M.C., Dose-response analyses of experimental cancer data, *Drug Metab. Rev.*, 32, 193, 2000.

Moschandreas, D.J. and Saksena, S., Modeling exposure to particulate matter, *Chemosphere*, 49, 1137, 2002.

Moya, J. and Phillips, L., Overview of the use of the U.S. EPA exposure factors handbook, *Int. J. Hyg. Environ. Health*, 205, 155, 2002.

Mueller, S.O., Xenoestrogens: mechanisms of action and detection methods, *Anal. Bioanal. Chem.*, 378, 582, 2004.

Palmer, J.R. et al., Risk of breast cancer in women exposed to diethylstilbestrol *in utero*: preliminary results (U.S.), *Cancer Causes Control*, 13, 753, 2002.

Pennington, D. et al., Assessing human health response in life cycle assessment using ED_{10}s and DALYs: part 2, noncancer effects, *Risk Anal.*, 22, 947, 2002.

Schneider, T. et al., Conceptual model for assessment of dermal exposure, *Occup. Environ. Med.*, 56, 765, 1999.

Strickland, J.A. and Foureman, G.L., U.S. EPA's acute reference exposure methodology for acute inhalation exposures, *Sci. Total Environ.*, 288, 51, 2002.

Tanaka, E., Toxicological interactions between alcohol and benzodiazepines, *J. Toxicol. Clin. Toxicol.*, 40, 69, 2002.

Wexler, P. and Phillips, S., Tools for clinical toxicology on the World Wide Web: review and scenario, *J. Toxicol. Clin. Toxicol.*, 40, 893, 2002.

3 Toxicokinetics

3.1 INTRODUCTION

3.1.1 RELATIONSHIP TO PHARMACOKINETICS

The study of drug disposition in physiological compartments is traditionally referred to as pharmacokinetics. Toxicokinetics includes the more appropriate study of exogenous compounds (xenobiotics), the adverse effects associated with their concentrations, their passages through the physiological compartments, and their ultimate fates. Compartmental toxicokinetics involves administration or exposure to a chemical, rapid equilibration to the central compartments (plasma and tissues), followed by distribution to peripheral compartments. Consequently, the principles of pharmacokinetics and toxicokinetics historically have been used interchangeably.

3.1.2 ONE-COMPARTMENT MODEL

A one-compartment model is the simplest representation of a system that describes the first-order extravascular absorption of a xenobiotic. The absorption rate is constant (k_a) and allows for entry into central compartments (plasma and tissues). A constant first-order elimination (k_{el}) is explained below. The reaction is illustrated below. Few compounds follow one-compartment modeling; agents such as creatinine equilibrate rapidly, distribute uniformly throughout blood, plasma, and tissues, and are eliminated quickly.

$$\text{Reaction 1: Chemical} \xrightarrow{\;k_a\;} \text{One-Compartment} \xrightarrow{\;k_{el}\;} \text{Excreted}$$

Drug or toxin elimination processes generally follow zero-order or first-order kinetics. *First-order elimination* and first-order kinetics terms describe the rate of elimination of a drug in proportion to the drug concentration: an increase in the concentration of the chemical results in an increase in the rate of elimination. This usually occurs at low drug concentrations. Alternatively, *zero-order elimination* refers to the removal of a fixed amount of drug independent of the concentration: increasing plasma concentration of the chemical does not proceed with a corresponding increase in elimination. Zero-order kinetics usually occurs when biotransforming enzymes are saturated. Clinically, this situation is the most common cause of drug accumulation and toxicity. *Michaelis-Menton kinetics* occurs when a compound is metabolized according to both processes. For instance, at low concentrations, ethanol is metabolized by first-order kinetics (dose-dependent). As the blood alcohol concentration (BAC) increases, the process switches to zero-order (dose-independent) elimination.

3.1.3 Two-Compartment Model

In a two-compartment model, a xenobiotic chemical requires a longer time for the concentration to equilibrate with tissues and plasma. Essentially, molecules enter the central compartment at a constant rate (k_a) and may also be eliminated through a constant first-order elimination (k_{el}) process. However, a parallel reaction (see below) occurs when the chemical enters and exits peripheral compartments through first-order rate constants for distribution (k_{d1} and k_{d2}, respectively). The distribution phases are variable but rapid, relative to elimination. Most drugs and xenobiotics undergo two- or multi-compartment modeling behavior.

$$\text{Reaction 2: Chemical} \xrightarrow{k_a} \text{First-Compartment} \xrightarrow{k_{el}} \text{Excreted}$$

$$k_{d1} \Updownarrow k_{d2}$$

$$\text{Second Compartment}$$

3.1.4 Applications to Toxicology Testing

Clinical applications of kinetics, and the phenomena of absorption, distribution, metabolism, and elimination (ADME) are useful in the monitoring of toxicological activities of chemicals. Assigning a compound to a compartmental system, particularly a two-compartment model or the more complex multi-compartment model, allows for prediction of adverse effects of agents based on distribution and equilibration within the system. ADME factors that influence the fate of a chemical are discussed below.

3.2 ABSORPTION

3.2.1 Ionic and Non-Ionic Principles

How does the ionic or non-ionic species of a chemical relate to absorption? The eukaryotic cell membrane is characterized as a flexible, semi-permeable barrier that maintains a homeostatic environment for normal cellular functions by preventing chemicals, ions, and water from traversing through it easily. Figure 3.1 illustrates the structure of a eukaryotic cell membrane. The membrane consists of phospholipid polar heads, glycolipids, integral proteins organized toward the periphery of the membrane (hydrophilic region), and non-polar tails and cholesterol directed inward (hydrophobic region). The rest of the membrane is interspersed with transmembrane channel proteins and other intrinsic proteins. The fluidity of the membrane is due to the presence of cholesterol and integral proteins. The phospholipid polar groups and non-polar tails are derived from triglycerides. Thus, the lipid bilayer structure of the cell membrane owes its functional activity in part to the polar environment

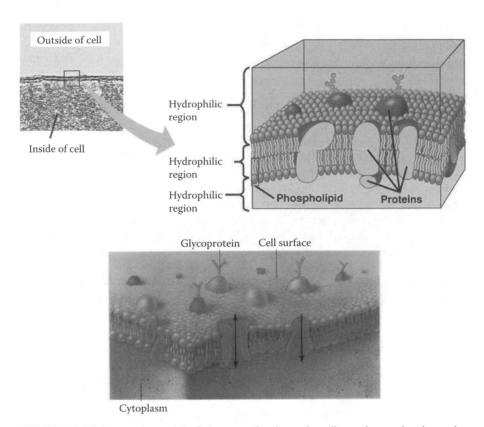

FIGURE 3.1 Fluid mosaic model of the normal eukaryotic cell membrane showing polar heads (hydrophilic regions) and non-polar tails (hydrophobic region). Other structural features, such as transmembrane proteins and surface glycoproteins, contribute to the fluidity and selective permeability of the membrane. (Photo courtesy of Mr. Trevor Gallant, http://kvhs.nbed.nb.ca/gallant/biology/biology.html. With permission.)

contributed by the phosphate and carboxyl moieties (heads) and the relatively non-polar saturated long carbon chains (tails).

Because of the selectively permeable arrangement of the cell membrane, the ability of a chemical to bind to or penetrate intact membranes is determined by its water or lipid solubility. In turn, the relative contribution of a chemical to its polarity is influenced by the acidic or basic nature of the chemical in solution, which is determined by the proportion of the chemical's propensity to dissociate or associate. According to the Bronsted-Lowry theory of acids and bases, an acidic compound dissociates (donates a proton); a basic molecule associates (accepts a proton). The dissociation constant for an acid can be summarized according to the following formula:

$$K_a = \frac{[H^+][A^-]}{[HA]}$$

where K_a is the dissociation constant for an acid; [H+], [A-], and [HA] represent the hydrogen ion, conjugate acid, and undissociated acid concentrations, respectively. The pK_a derived from the negative logarithm of the acid dissociation constant is calculated as follows:

$$pK_a = -\log K_a$$

The pK_a of a drug, therefore, is defined as the pH at which 50% of the compound is ionized. Similarly, pK_b is the negative logarithm of the base dissociation constant (K_b). In general, acids and bases are classified according to their abilities to dissociate relative to the pH of the solution. Independent of the pH, however, the lower the pK_a of an acid, the stronger the acid. Similarly, the lower the pK_a (or higher the pK_b) of a base, the stronger the base. Unlike most chemicals that display a wide variety of dissociative abilities in solution, therapeutic drugs are usually weak acids or bases, neutral, or amphoteric compounds.

In general, strong acidic and basic characteristics of chemicals are determined by the relative presence of carboxyl anions and amine groups in solution, respectively. For instance, benzoic acid, with a pK_a of 4, readily dissociates, leaving the carboxyl anion after dissociation. Its low pK_a calculates a high dissociation constant (K_a) equal to 1×10^{-4}, resulting in a compound possessing strong acidic properties. Similarly, aniline is a highly protonated species with a pK_b of 10 and corresponding K_b of 1×10^{-10} ($pK_a + pK_b = 14$; thus, although it is a base possessing an amine group, the pK_a of aniline can also be expressed as 4). The low dissociation constant of the base indicates that the H+ ions are closely held to the nitrogen of the amine and contribute to its strong basic nature.

3.2.2 HENDERSON-HASSELBACH EQUATION

The relationship of the ionization of a weakly acidic drug, its pK_a, and the pH of a solution can be shown by the Henderson-Hasselbach equation. The equation allows for the prediction of the non-ionic and ionic states of a compound at a given pH. The formulas are:

$$\text{Acid: pH} = pK_a + \log ([A^-]/[HA])$$

$$\text{Base: pH} = pK_a + \log ([HA]/[A^-])$$

The equations are derived from the logarithmic expression of the dissociation constant formula above. Small changes in pH near the pK_a of a weakly acidic or basic drug markedly affect its degree of ionization. This is more clearly shown with rearrangement of the Henderson-Hasselbach equations, such that:

$$\text{For an acid: } pK_a - pH = \log ([HA]/[A^-])$$

$$\text{For a base: } pK_a - pH = \log ([A^-]/[HA])$$

TABLE 3.1
Summary of Chemical Properties and Behavior of Acidic, Neutral, and Basic Compounds in the Stomach Environment (pH = 2)

Compound Properties	Acidic	Neutral (Weakly Acidic)	Basic
pKa	3	7	10
Non-ionic:ionic ratio	10:1	10^5:1	1:10^8
Acidic/basic nature	Free acid, highly lipophilic	Non-ionic, lipophilic	Protonated, hydrophilic, very low lipophilicity
Absorption	Favorable	Favorable	Not favorable

For an acidic compound present in the stomach (e.g., $pK_a = 4$ in an average pH of the stomach = 2), inserting the numbers into the Henderson-Hasselbach equation results in a relative ratio of non-ionized to ionized species of 100:1. This transforms the compound predominantly to the non-ionic form within the acidic environment of the stomach, rendering it more lipophilic. Lipophilic compounds have a greater tendency for absorption within that compartment.*

In the proximal small intestine areas of the duodenum and jejunum where the pH is approximately 8, the same compound will be predominantly in the ionized state. The relative ratio of non-ionized to ionized species is reversed (1:10^4). Thus, within the weakly basic environment of the proximal intestine, a strongly acidic drug is less lipophilic, more ionized, and slower to be absorbed.

Conversely, for a strong basic compound with a $pK_a = 4$ ($pK_b = 10$) in the stomach, the Henderson-Hasselbach equation predicts that the ratio of ionized to non-ionized species equals 100:1. Thus, a basic compound is more ionized, less lipophilic, and slower to be absorbed in the stomach. In the basic environment of the proximal intestine (pH = 8), however, the ionized to non-ionized species ratio is 1:10^4, rendering it more lipophilic and imparting a greater propensity for absorption.

The knowledge of the pK_a of a chemical is useful in predicting its absorption in these compartments. In addition, as explained below, this information is helpful in determining distribution and elimination functions. Tables 3.1 and 3.2 summarize the chemical properties and behaviors of acidic, neutral, and basic chemicals and their ionization potential in the stomach and small intestine. Based on the extremely acidic environment of an empty stomach, acids or bases are either completely non-ionized and highly lipophilic or ionized and hydrophilic. In the basic environment of the small intestine, although some ionization is present, absorption of a neutral compound (weak acid) is favored over that of an acid.

Note that some toxicokinetic principles that govern chemical absorption are not always significant factors in the determination of toxic effects. This is primarily due to the circumstances surrounding toxic chemical exposure. Consequently, some of

---

* Note that the absorption rate of the stomach is limited and is secondary to its digestion and churning functions.

TABLE 3.2
Chemical Properties and Behaviors of Acidic, Neutral, and Basic
Compounds in the Small Intestine (pH = 8 to 10)

Compound Properties	Acidic	Neutral (Weakly Acidic)	Basic
pKa	3	7	10
Non-ionic:ionic ratio	$10^{-5}:1$ to $10^{-7}:1$	$10^{-1}:1$ to $10^{-3}:1$	$10^{-2}:1$ to $1:1$
Acidic/basic nature	Proton donor, very low lipophilic	Weak acid, some ionization, low lipophilicity	Some ionization, lipophilic
Absorption	Not favorable	Some absorption over length of intestinal tract	Favorable

the circumstances that influence chemical absorption in a therapeutic setting will not be considered here. These factors include formulation and physical characteristics of a drug product; drug interactions; presence of food in the intestinal tract; gastric emptying time; and concurrent presence of gastrointestinal diseases. Individual categories of toxic exposure and special circumstances that alter their effects will be considered in later chapters.

3.2.3 ABSORPTION IN NASAL AND RESPIRATORY MUCOSA

3.2.3.1 Nasal Mucosa

The vestibule of the nasal cavity and nasopharynx is covered by olfactory epithelium. This mucous membrane lining consists of an extensive network of capillaries and pseudostratified columnar epithelium interspersed with goblet cells. Inspired air enters the nasal conchae and is warmed by circulating blood in the capillaries, while mucus secreted by the goblet cells moistens the air and traps airborne particles. Thus, several factors facilitate rapid availability and absorption from chemical exposure through the nasal route:

1. Vast surface areas provided by capillary circulation and epithelial cells in mucous membranes
2. Ability of the nasal mucosa to trap dissolved and particulate substances
3. Phospholipid secretions of the nasal mucosa (resulting in a relatively neutral pH)

Non-ionized drugs are more rapidly absorbed in this compartment relative to their ionized counterparts.

3.2.3.2 Respiratory Mucosa

The upper and lower respiratory tracts extend from the nasal cavity to the lungs. The conducting portion responsible for ventilation consists of the pharynx, larynx, trachea, bronchi, bronchioles, and terminal bronchioles. The respiratory portion that serves as the main site for gas exchange consists of the respiratory bronchioles, alveolar ducts, alveolar sacs, and alveoli. Most of the upper and lower tracts are covered by ciliated secretory epithelial cells.

Aside from the nasal epithelium, however, significant absorption of airborne substances does not occur above the level of the respiratory bronchioles. As respiratory bronchioles penetrate deeply into the lungs, the epithelial lining changes from simple cuboidal to simple squamous, enhancing the surface absorptive capability of the mucosa. In addition, the thin 0.5-μm respiratory membrane is composed of 300 million alveoli and further supported by an extensive network of capillary endothelium (see Figure 2.2, Chapter 2). This anatomy allows rapid diffusion of chemicals released as gases within aerosols or carried by humidified air.

3.2.4 Transport of Molecules

Absorption of chemicals requires basic mechanisms of transporting molecules across cell membranes and fluid barriers. These transport vehicles include diffusion, active transport, facilitated diffusion, and filtration.

The simplest mechanism for transportation of molecules involves diffusion, defined as the transport of molecules across a semi-permeable membrane. The most common pathway of diffusion is passive transport by which molecules are transported from an area of high to low solute concentration, down the concentration gradient. This is not an energy-dependent process, and no electrical gradient is generated. Lipophilic molecules, small ions, and electrolytes generally gain access through membrane compartments by passive diffusion because they are not repelled by the phospholipid bilayers of cell membranes. Most passive diffusion processes, however, are not molecularly selective.

In contrast to passive transport, polar substances such as amino acids, nucleic acids, and carbohydrates are moved by an active transport process. This mechanism shuttles ionic substances from areas of low solute concentration to high solute concentration, against a concentration gradient. Active transport is an energy-dependent process and requires macromolecular carriers. These two requirements render this process susceptible to metabolic inhibition and consequently a site for toxicological or pharmacological influence. In addition, active transport processes are saturable systems, implying that they have a limited number of carriers, making them competitive and selective.

Facilitated diffusion resembles active transport with the exception that substrate movement is not compulsory against a concentration gradient and it is not energy-dependent. For example, macromolecular carriers augment the transfer of glucose across gastrointestinal epithelium through to the underlying capillary circulation.

Filtration allows for the movement of solutes across epithelial membrane barriers along with water movement. The motive force here is hydrostatic pressure. In general, the molecular weight cutoff for solutes transported along with water is about 100 to 200 mw units. Under normal circumstances, several membranes throughout the body with pores or fenestrations averaging 4 Å in diameter are designed to allow entry for larger molecules. These special membranes permit coarse filtration of lipophilic or hydrophilic substances. Such filtration processes are described for the renal nephron. The filtration slits formed by the interdigitating podocytes in the glomerular apparatus and Bowman's capsule prevent the passage of cells and solutes above 45 kD. Under normal circumstances, most smaller molecules pass through the filter, only to be reabsorbed in the proximal and distal convoluted tubules and collecting tubules.

Many organs exhibit passive filtration as a fundamental process for the passage of water. Intestinal absorption of water involves a transport method by which water and dissolved nutrients are coarsely filtered and absorbed through the intestinal lumen. At the tissue level, the inotropic action of the cardiac system is the driving force for capillary perfusion. The thin endothelial and epithelial membranes that form the walls of the pulmonary alveoli represent the only barriers for the diffusion of gases into the alveolar spaces.

3.3 DISTRIBUTION

3.3.1 Fluid Compartments

Anatomically, the body is not a homogeneous entity. Since approximately 70% of total body weight is composed of water, the physiological water compartments play significant roles in chemical distribution and its ultimate fate. Table 3.3 illustrates the composition and contribution of water-based compartments relative to body weight. The ability of a compound to gain access to these compartments is determined by a number of factors that influence the compound's subsequent fate. The distribution of the substance to the target organ (receptor site) and as to the sites of metabolism and excretion is reasonably predicted by such factors. These include: (1) physicochemical properties of a substance such as lipid solubility and ionic or non-ionic principles; (2) binding to macromolecules; and (3) physiological forces of blood flow and capillary perfusion. The cooperation of theses forces determines the relative bioavailability of a compound. These concepts are further discussed below.

TABLE 3.3
Compartments and Their Relative Percentages of Total Body Weight

Compartment	Description	Relative Percent of Body Weight
Plasma water (PL)	Non-cellular water component of blood	10
Interstitial water (IS)	Fluids relating to the interstices (spaces) of tissue	15
Extracellular water (EC)	PL + IS	25
Intracellular water (IC)	Confined to cytosol (within cells)	45
Total body water (TBW)	EC + IC	70
Solids	Tissue binding capacity	30
Total body weight	Solids + TBW	100

3.3.2 IONIC AND NON-IONIC PRINCIPLES

The same mechanisms that govern absorption of molecules across semi-permeable membranes also determine the bioavailability of a compound. Thus a compound with a propensity for a non-ionic existence is more likely to be lipophilic and thus has greater ability to penetrate phospholipid membranes and greater access for distribution to liver, kidney, muscle, adipose, and other tissues. Conversely, molecules that are highly ionic in plasma have diminished access to organs and tissues. As with absorption, pH and pK_a have similar influences over ionization or the lack of ionization at the tissue level. Similarly, as an indicator of distribution to organs, the apparent volume of distribution (V_d) is used as a measure of the amount of chemical in the entire body (mg/kg) relative to that in the plasma (mg/L) over a specified period. Rearranging the formula defines the relationship as:

$$V_d \text{ (L/kg)} = \text{dose (mg/kg)/plasma concentration (mg/L)}$$

Thus, V_d is useful in estimating the extent of distribution of a substance in the body. Table 3.4 shows the V_d values for 25 chemicals whose toxicity is well established. Examination of these numbers indicates that a compound with a large V_d (e.g., 19.0, d-propoxyphene HCl) is highly lipophilic and sequesters throughout tissue compartments, especially the central nervous system. V_d necessarily demonstrates a low plasma concentration relative to dose. Conversely, a compound with a low V_d (0.4, caffeine) is highly bound to circulating plasma proteins and thus has a plasma concentration 2.5-fold of the mean total body dose.

Another criterion for estimating bioavailability is the calculation of the area under the plasma concentration versus time curve (shaded area, AUC, Figure 3.2). Dividing AUC after an oral dose by AUC after an intravenous dose and multiplying by 100 yields percent bioavailability. This value is used as an indicator for the fraction of chemical absorbed from the intestinal tract that reaches the systemic circulation after a specified period. As is explained below, other factors such as first pass metabolism, rapid protein uptake, and rapid clearance may alter the calculation. The ability to distribute to target tissues is then determined by the V_d.

3.3.3 PLASMA PROTEIN BINDING

Polar amino acids projecting from the surfaces of circulating proteins induce non-specific, reversible binding of many structurally diverse small molecules. Circulating proteins transport chemicals through the circulation to target receptor sites. Such proteins include serum albumin; hemoglobin that binds iron; $\alpha 1$ globulins such as transcortin that bind steroid hormones and thyroxin; $\alpha 2$ globulins such as ceruloplasmin that bind various ions; $\beta 1$ globulins such as transferrin that bind iron; lipoproteins; and red blood cells. Factors that influence the extent of protein binding are dictated by non-ionic and ionic forces of radicals. Most binding is reversible and displaceable, primarily because it involves non-covalent, hydrogen, ionic, or van der Waals forces. The binding renders molecules non-absorbable and unable to bind to target receptor sites.

TABLE 3.4
Human V_d Values of 25 Representative Chemicals and Drugs with Known Toxicities

Chemical	V_d
Acetylsalicylate	0.2
Diazepam	1.65
Digoxin	6.25
Methanol	0.6
Isopropanol	0.6
Phenol	1.0
Na fluoride	0.6
Malathion	1.0
Nicotine	1.0
Lithium SO_4	0.9
d-Propoxyphene HCl	19.0
Phenobarbital	0.55
Arsenic O_3	0.2
Mercuric Cl_2	1.0
Thallium SO_4	1.0
Lindane	1.0
Carbon tetrachloride	1.0
Dichloromethane	1.0
Hexachlorophene	1.0
Verapamil HCl	4.5
Orphenadrine HCl	6.1
Fenytoin	0.65
Na oxalate	1.0
Caffeine	0.4
K chloride	1.0

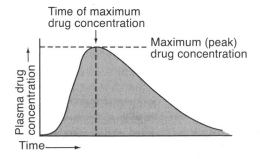

FIGURE 3.2 Plasma concentration versus time curve for drug X. Shaded region represents area under the curve (AUC). (From Barile, Frank A., *Clinical Toxicology: Principles and Mechanisms,* CRC Press, 2004.)

Serum albumin poses particularly significant alteration of chemical and drug toxicokinetics and pharmacokinetics by virtue of its high affinity for acidic drugs such as warfarin. The binding of a substance to a circulating transport protein forms a chemical–protein complex such that its activity, metabolism, and rate of elimination are determined by the strength of this bond and diffusion into various compartments. The rate at which a chemical–protein complex dissociates, therefore, determines its toxicological, pharmacological, and metabolic fate.

The plasma concentration of a chemical is driven to equilibrium by the rate of cardiac output, and, assuming steady-state conditions, reaches a plateau concentration. In addition, consider that only unbound chemicals diffuse into tissues for access to target sites, receptors, or for biotransformation because a chemical–protein complex cannot pass through semi-permeable cell membranes. As an unbound transformed molecule is excreted, the dissociation of the chemical–protein complex increases, shifting the equilibrium toward the unbound chemical. This shift continues until the free chemical concentration is equal to that in plasma and tissues (i.e., steady state). Thus, this complex acts as a depot, releasing enough active ingredient as determined by its chemical properties.

Chemicals that are extensively bound to circulating or tissue proteins pose significant toxicological risks. Toxic effects of such chemicals extend well beyond acute exposures. Chronic and subchronic exposure to lead, for instance, has significant long-term neurological consequences by virtue of its affinity for depositing in bone. Therapeutically, steady-state concentrations of warfarin are easily disrupted when another acidic drug such as acetylsalicylic acid (aspirin) displaces it from circulating albumin binding sites. Since warfarin is extensively protein bound (97%), a displacement of as little as 1% can double the amount of warfarin available for interaction with target organs, resulting in significant adverse hemodynamic events. Consequently, displacement of protein-bound chemicals results in increased unbound chemical concentration in plasma and increased toxicological and pharmacological effects, followed by enhanced metabolism.*

3.3.4 Lipids

Distribution of chemicals is also determined in part by their lipid/water partition coefficients. Two chemicals with similar pK_a values such as amphetamine (9.8) and chlorpromazine (9.3) have different lipid affinities in their non-ionized forms. At basic pH (8 to 9), both chemicals are completely non-ionized. However, at pH 10 or above, the partition coefficient of chlorpromazine in lipid medium ($K_d = 10^{6.3}$) is $10^{3.5}$ times greater than that of amphetamine ($K_d = 10^{1.8}$), allowing for greater absorption of the former in the contents of the intestinal tract. Similarly, the polychlorinated biphenyls (PCBs) and the DDT contact insecticide have high lipid solubilities, dissolve in neutral fats, and thus readily accumulate in adipose tissue.

---

* It is important to note that availability of chemical at a receptor site is influenced more by rate of diffusion (V_d) than by rate of dissociation. This is especially true when the displacement of an acidic chemical with a small apparent V_d displays toxicologically important consequences.

3.3.5 Liver and Kidney

The high capacities of the liver and kidney for binding and concentrating toxicants coupled with their extensive circulation and metabolic activities render these organs susceptible to chemical attack. Further discussion of the metabolic roles that the liver and kidney play in chemical detoxification are addressed below.

3.3.6 Blood–Brain Barrier

Formed by the tight junctions of capillary endothelial cells and surrounded by basement membrane and an additional layer of neuroglial cells, the blood–brain barrier (BBB) imposes a formidable hindrance to the passage of molecules. Although often misconstrued as an anatomical structure, it is primarily a physiologically functional entity, selectively allowing passage of a few water soluble molecules, ions, and glucose (by active transport). It is sufficiently impermeable to most compounds including cells, proteins, and chemicals. Lipid-soluble chemicals such as oxygen, carbon dioxide, alcohol, anesthetics, and the opioid analgesics enjoy relatively easy access to the central nervous system via the intact BBB. When the BBB loses its barrier function as occurs with bacterial or viral encephalitis, the compromised membrane is then susceptible to unregulated entry.

3.3.7 Placenta

The placenta is fully developed by the third month of pregnancy and is formed by the embryonic chorion (syncytial trophoblast) and the maternal endometrium (decidua basalis). Although gases, vitamins, amino acids, small sugars, ions, and waste products diffuse across or are stored in the placental membrane, viruses and lipid soluble chemicals have relative easy access to the fetus. Thus, the same principles of ionization that govern absorption of chemicals influence distribution and passage through the placenta. Non-ionized lipid-soluble substances penetrate more easily than ionized water-soluble materials. In addition, passage through to fetal membranes does not depend on the number of placental layers formed as a fetus matures. Biotransformation properties of the placenta are addressed below.

3.4 BIOTRANSFORMATION

3.4.1 Principles of Detoxification

Biotransformation is defined as the alteration of a chemical (xenobiotic) in a biological system.* Xenobiotic transformation plays a crucial role in maintaining homeostasis during chemical exposure. This is accomplished by converting lipid-soluble (non-polar) non-excretable xenobiotics to polar water-soluble compounds accessible for elimination in the bile and urine. The outcomes of biotransformation include facilitation of excretion, conversion of toxic parent compounds to non-toxic

---

* Exogenous chemicals are often referred to as xenobiotics when foreign substances access biological systems.

metabolites (detoxification), and conversion of non-toxic or non-reactive parent compounds to toxic or reactive metabolites (bioactivation).

3.4.2 BIOCHEMICAL PATHWAYS

Catalysts for xenobiotic transformation are incorporated into phase I and phase II reactions. Although occurring primarily in the liver, other organs such as kidneys, lungs, and dermal tissues have large capacities for these reactions.

Phase I reactions involve hydrolysis, reduction, and oxidation of chemicals to more hydrophilic, usually smaller, entities. Phase II reactions follow with glucuronidation, sulfation, acetylation, methylation, and conjugation with amino acids of the phase I metabolites. Phase II enzymes act on hydroxyl, sulfhydril, amino, and carboxyl functional groups present on either the parent compound or the phase I metabolite. Therefore, if a functional group is present on the parent compound, phase I reactions are circumvented.*

3.4.3 ENZYME SYSTEMS

3.4.3.1 P450 Enzyme Pathways

Among phase I biotransformation enzymes, the cytochrome P450 system is the most versatile and ubiquitous. The enzymes are predominantly mono-oxygenase, heme-containing proteins, found mostly in liver microsomes and mitochondria. The enzymes are classified into families, subfamilies, or subtypes of subfamilies, according to their amino acid sequences, and they catalyze the incorporation of one atom of oxygen or facilitate the removal of one atom of hydrogen in the substrate. Other types of enzymes involved in the P450 system are hydroxylases, dehydrogenases, esterases, and N-dealkylation and oxygen transferases.

Phase II biotransformation enzymes include UDP-glucuronosyl transferase, sulfotransferases, acetyltransferases, methyltransferases, and glutathione S-transferases. Except for acetylation and methylation reactions, phase II reactions along with enzyme cofactors interact with different functional groups to increase hydrophilic properties. Glucuronidation is the main pathway for phase II reactions. It converts non-polar lipid-soluble compounds to polar water soluble metabolites. Sulfation reactions involve the transfer of sulfate groups from PAPS (3-phosphoadenosine-5-phosphosulfate) to the xenobiotic, yielding sulfuric acid esters that are eliminated into the urine.

3.4.3.2 Glutathione Conjugation

Glutathione (GSH) conjugation is responsible for detoxification (or bioactivation) of xenobiotics. GSH conjugates are excreted intact in bile or converted to mercapturic acid in the kidneys and eliminated in the urine.

---

* Because of the presence of hydroxyl groups, morphine undergoes direct phase II conversion to morphine-3-glucuronide, whereas codeine must endure transformation with both phase I and II reactions.

3.4.3.3   Other Oxidative–Reductive Reaction Systems

Enzymes used in hydrolysis phase I reactions include carboxylesterase, azo- and nitro-reductases, and quinone reductases. Carboxylesterases cleave carboxylic acid esters, amides, and thioesters. Azo- and nitro- reduction reactions incorporate two enzymes: azo- (–N=N–) reductases from intestinal flora and liver cytochrome P450 reductases. Quinone reduction reactions involve quinone reductases and NADPH cytochrome P450 reductases. The former are involved in two-electron reduction reactions (detoxification), and the latter in one-electron reduction reactions (bioactivation).

Oxidative enzymes include alcohol dehydrogenases and cytochrome P450 oxidases. For example, alcohol dehydrogenase (ADH) and aldehyde dehydrogenases convert alcohols to corresponding aldehydes and acids, respectively.

3.5   ELIMINATION

3.5.1   Urinary Excretion

Urinary excretion represents the most important route of chemical, drug, and waste elimination in the body. Normal glomerular filtration rate (GFR, ~110 to 125 ml/min) is determined by the endothelial fenestrations (pores) formed by the visceral layers of the specialized endothelial cells of the glomeruli. The pores are coupled with the overlying glomerular basement membranes and the filtration slits of the pedicels. This three-layer membrane permits the passage of non-cellular materials, with a molecular weight cutoff of ~45 to 60K. The filtrate is then collected in Bowman's capsule. Reabsorption follows in the proximal convoluted tubules (PCTs), loop of Henle, and distal convoluted tubules (DCTs). The collecting ducts have limited reabsorption and secretion capabilities. Control of the transfer of molecules back to the capillary circulation or elimination in the final filtrate depends predominantly on reabsorption and secretion processes.

As with absorption through biological membranes and distribution in physiological compartments, urinary elimination of toxicants is guided by ionic and nonionic principles. While Na^+ and glucose are actively transported or facilitated by symporters across the proximal tubules, other larger polar compounds are prevented from reabsorption to the capillary circulation and are eventually eliminated in the urine. Non-ionic non-polar compounds are transported by virtue of their ability to move across phospholipid membranes. Clinically, these principles are somewhat susceptible to manipulation for reducing chemical toxicity. For instance, it has long been proposed that treatment of overdoses of barbiturates (pK_a values between 7.2 and 7.5) may be effectively treated with sodium bicarbonate. The mechanism of this therapeutic intervention relies, in theory, on the alkalinization of the urine, rendering the acidic drug ionic, water soluble, and more amenable to elimination. In practice, however, the buffering capacity of the plasma coupled with a large dose of the sedative or hypnotic makes this a treatment modality of last resort.

Other organic ions undergo active tubular secretion. Organic anion transporter (oat) proteins (organic acid transporters) and organic cation transporter (oct) proteins (organic base transporters) both actively secrete polar molecules from capillaries to

TABLE 3.5
Bile-to-Plasma Ratios of Representative Xenobiotics

Classification	Bile-to-plasma Ratio	Representative Compounds
Class A	1 to 10	Na+, K+, glucose
Class B	10 to 1000	Arsenic, bile acids, lidocaine, phenobarbital
Class C	Less than 1	Albumin, metals (zinc, iron, gold, chromium)

the PCT lumen. Some classic examples include the active tubular secretion of para-aminohippuric acid (PAH) by oat proteins as well as sulfonamide antibiotics and methotrexate that compete for the same secretory pathways. Probenecid, a uricosuric agent used in the treatment of gout, competes for oat proteins as part of its therapeutic mechanism. While small doses of the drug actually depress the excretion of uric acid by interfering with tubular secretion but not reabsorption, larger doses depress the reabsorption of uric acid and lead to increased excretion and decreases in serum levels. Similarly, small doses of probenecid decrease the renal excretion of penicillins that compete with the same oat proteins, necessitating adjustment of antibiotic dosage regimens.

3.5.2 FECAL ELIMINATION

With the exception of cholestyramine, a cholesterol-binding drug, and the herbicide paraquat, the presence of non-absorbable, non-transformed xenobiotics in the feces is rare. This is because many metabolites of compounds formed in the liver are excreted into the intestinal tract via the bile. Furthermore, organic anions and cations are biotransformed and actively transported into bile by carrier mechanisms similar to those in the renal tubules.

Chemicals ingested via oral exposure enter the hepatic portal system and liver sinusoids, only to be extensively cleared by the liver. This toxicokinetic phenomenon is known as hepatic first pass elimination. According to the principles of elimination, compounds may be classified based on their bile-to-plasma (B:P) ratios. Table 3.5 outlines the classifications of representative chemicals based on this ratio. Substances with high B:P ratios are readily eliminated in bile. Drugs such as phenobarbital are extensively removed by the liver, which explains why the oral bioavailability of these drugs is low despite complete absorption. Other compounds, such as lidocaine, a local anesthetic, are extensively eliminated by first pass elimination and thus oral administration is precluded. Compounds with low B:P values are not subject to this phenomenon and maintain relatively higher plasma and other compartmental concentrations.

3.5.3 PULMONARY ELIMINATION

Most gases, aerosolized molecules, and volatile liquids are eliminated by simple diffusion via exhalation. The rate of pulmonary elimination is inversely proportional to the blood solubility of a compound. Thus, compounds with low blood solubility, such as ethylene dioxide, have high vapor pressure levels and are readily volatilized and eliminated as pulmonary effluents. Conversely, compounds with high blood solubility levels, such as ethanol, carbon tetrachloride, and chloroform, have low vapor pressures. Their elimination through the lungs is more a function of concentration than of solubility. Similarly, lipid-soluble general anesthetic agents have long half-lives in plasma because of their low vapor pressures and high plasma solubilities.

3.5.4 MAMMARY GLAND SECRETION

The mammary glands are modified sudoriferous glands that produce a lipid-laden secretion (breast milk) that is excreted during lactation. The lobules of the glands consist of alveoli surrounded by one or two layers of epithelial and myoepithelial cells anchored to well-developed basement membranes. Unlike the BBB and placenta, these cells do not interpose a significant barrier to transport of molecules. In addition, breast milk is a watery secretion with a significant lipid component. Consequently, both water- and lipid-soluble substances are transferred through the breast milk. In particular, basic chemicals readily accumulate in the weakly acidic or alkaline environment of breast milk (pH 6.5 to 7.5). Accordingly, nursing infants are exposed to almost any substance present in the maternal circulation.

3.5.5 OTHER SECRETIONS

Elimination of chemicals through saliva, tears, hair follicles, and sudoriferous (watery) or sebaceous (oily) sweat glands is a function of plasma concentration. Elimination is also influenced by liver metabolism, passage through the intestinal tract and pulmonary system, and by limited access to local glandular secretions. Thus, secretion of compounds through mucous, ophthalmic, and dermal glands is not significant. In forensic toxicology, however, certain chemicals may be detected in minute concentrations in hair follicles and nails when plasma levels seem to be nonexistent. Chronic arsenic and lead poisoning has been detected by measurement of the metals bound to hair follicles and hair shafts.

SUGGESTED READINGS

Kauffman, F.C., Sulfonation in pharmacology and toxicology, *Drug Metab Rev.*, 36, 823, 2004.

Klaassen, C.D. and Slitt, A.L., Regulation of hepatic transporters by xenobiotic receptors, *Curr. Drug Metab.*, 6, 309, 2005.

Medinsky, M.A. and Valentine, J.L., Principles of toxicology, in *Casarett and Doull's Toxicology: The Basic Science of Poisons*, Klaassen, C.D., Ed., 6th ed., McGraw-Hill, New York, 2001, Ch. 7.

Norinder, U. and Bergstrom, C.A., Prediction of ADMET properties, *Chem. Med. Chem.*1, 920, 2006.

Rodriguez-Antona, C. and Ingelman-Sundberg, M., Cytochrome P450 pharmacogenetics and cancer, *Oncogene*, 25, 1679, 2006.

Strolin-Benedetti, M., Whomsley, R., and Baltes, E.L., Differences in absorption, distribution, metabolism and excretion of xenobiotics between the paediatric and adult populations, *Expert Opin. Drug. Metab. Toxicol.,* 1, 447, 2005.

Urso, R., Blardi, P., and Giorgi, G., A short introduction to pharmacokinetics, *Eur. Rev. Med. Pharmacol. Sci.,* 6, 33, 2002.

Wijnand, H.P., Pharmacokinetic model equations for the one- and two-compartment models with first-order processes in which the absorption and exponential elimination or distribution rate constants are equal, *J. Pharmacokinet. Biopharm.,* 16, 109, 1988.

REVIEW ARTICLES

Barile, F.A., Dierickx, P.J., and Kristen, U., *In vitro* cytotoxicity testing for prediction of acute human toxicity, *Cell Biol. Toxicol.,* 10, 155, 1994.

Bodo, A. et al., The role of multidrug transporters in drug availability, metabolism and toxicity, *Toxicol. Lett.,* 140, 133, 2003.

Buchanan, J.R., Burka, L.T., and Melnick, R.L., Purpose and guidelines for toxicokinetic studies within the National Toxicology Program, *Environ. Health Perspect.,* 105, 468, 1997.

Cascorbi, I., Genetic basis of toxic reactions to drugs and chemicals, *Toxicol. Lett.,* 162, 16, Epub 2005

Dalvie, D., Recent advances in the applications of radioisotopes in drug metabolism, toxicology and pharmacokinetics, *Curr. Pharm. Des.,* 6, 1009, 2000.

El-Tahtawy, A.A. et al., Evaluation of bioequivalence of highly variable drugs using clinical trial simulations II. Comparison of single and multiple-dose trials using AUC and Cmax, *Pharm. Res.,* 15, 98, 1998.

Gamage, N. et al. Human sulfotransferases and their role in chemical metabolism, *Toxicol Sci.* 90, 5, Epub 2005.

Ikeda, M., Application of biological monitoring to the diagnosis of poisoning, *J. Toxicol. Clin. Toxicol.,* 33, 617, 1995.

Ioannides, C. and Lewis, D.F., Cytochromes P450 in the bioactivation of chemicals, *Curr. Top. Med. Chem.,* 4, 1767, 2004.

Ishii, K. et al., A new pharmacokinetic model including *in vivo* dissolution and gastrointestinal transit parameters, *Biol. Pharm. Bull.,* 18, 882, 1995.

Krzyzanski, W. and Jusko, W.J., Mathematical formalism for the properties of four basic models of indirect pharmacodynamic responses, *J. Pharmacokinet. Biopharm.,* 25, 107, 1997.

Lacey, L.F. et al., Evaluation of different indirect measures of rate of drug absorption in comparative pharmacokinetic studies, *J. Pharm. Sci.,* 83, 212, 1994.

Lin, J.H., Species similarities and differences in pharmacokinetics, *Drug Metab. Dispos.,* 23, 1008, 1995.

Lock, E.A. and Smith, L.L., The role of mode of action studies in extrapolating to human risks in toxicology, *Toxicol. Lett.,* 140, 317, 2003.

Mahmood, I. and Miller, R., Comparison of the Bayesian approach and a limited sampling model for the estimation of AUC and Cmax: a computer simulation analysis, *Int. J. Clin. Pharmacol. Ther.,* 37, 439, 1999.

Matthies, M., Exposure assessment of environmental organic chemicals at contaminated sites: a multicompartment modelling approach, *Toxicol. Lett.,* 140, 367, 2003.

Meineke, I. and Gleiter, C.H., Assessment of drug accumulation in the evaluation of pharmacokinetic data, *J. Clin. Pharmacol.*, 38, 680, 1998.

Myllynen, P., Pasanen, M., and Pelkonen, O., Human placenta: a human organ for developmental toxicology research and biomonitoring, *Placenta*, 26, 361, 2005.

Perri, D. et al., The kidney: the body's playground for drugs: an overview of renal drug handling with selected clinical correlates, *Can. J. Clin. Pharmacol.*, 10, 17, 2003.

Plusquellec, Y., Courbon, F., and Nogarede, S., Consequence of equal absorption, distribution and/or elimination rate constants, *Eur. J. Drug Metab. Pharmacokinet.*, 24, 197, 1999.

Poulin, P. and Theil, F.P., Prediction of pharmacokinetics prior to *in vivo* studies II. Generic physiologically based pharmacokinetic models of drug disposition, *J. Pharm. Sci.*, 91, 1358, 2002.

Ramanathan, M., Pharmacokinetic variability and therapeutic drug monitoring actions at steady state, *Pharm. Res.*, 17, 589, 2000.

Repetto, M.R. and Repetto, M., Concentrations in human fluids: 101 drugs affecting the digestive system and metabolism, *J. Toxicol. Clin. Toxicol.*, 37, 1, 1999.

Rogers, J.F., Nafziger, A.N., and Bertino, J.S., Jr., Pharmacogenetics affects dosing, efficacy, and toxicity of cytochrome P450-metabolized drugs, *Am. J. Med.*, 113, 746, 2002.

4 Risk Assessment and Regulatory Toxicology

4.1 RISK ASSESSMENT

Risk assessment is a complex process aimed at determining, evaluating, and predicting the potential for harm to a defined community from exposure to chemicals and their by-products. The tolerance of chemicals in the public domain is initially determined by the need for such chemicals, by the availability of alternative methods for obtaining the desired qualities of the chemicals, and by the economic impact of the presence or absence of the chemical agents on the economic status of the region.

For example, over the last few decades, the amounts and frequencies of use of pesticides and herbicides have increased exponentially, especially in the agricultural and commercial industries. Use of these chemicals resulted in significant benefits for commercial, household, and industrial concerns, especially in relation to greater yields of agricultural products as a result of reduced encroachment by insects and wild-type plant growth. Indeed, greater availability of food products and more sustainable produce have been achieved. The balance between safe and effective use of such chemicals and the risk to the population constantly undergoes reevaluation, however, to determine whether pesticides and herbicides are associated with developmental, carcinogenic, endocrine, or other toxicologic abnormalities.

Quantitative risk assessment is an estimation of toxic exposure concentrations or doses that, within a defined probability level, lead to specific increases in lifetime incidence rates for an undesirable consequence associated with the toxic substance. Factors used to estimate prediction models depend largely on accumulated retrospective and prospective data including the applicability of the scientific knowledge. In actuality, risk assessment is based on a *lack* of available data for a substance with potential for human or animal toxicity and the calculation of an estimate of potential toxicity on the data that are available. Consequently, gathering reliable and valid toxicity data for the prediction of human and animal health risks relies on a systematic approach in the quantitative risk assessment evaluation process, including:

1. Hazard identification
2. Dose–response relationship
3. Potential for exposure
4. Characterization of risk

4.1.1 HAZARD IDENTIFICATION

In an effort to determine whether a health risk exists, the process of hazard identification uses qualitative methods to evaluate the potential for adverse health effects of chemical or physical substances. The information incorporated into the database is obtained from relevant retrospective epidemiological studies and case studies. In addition, controlled animal and *in vitro* studies (including teratogenicity, mutagenicity, and carcinogenicity studies) are also conducted to contribute to risk assessment evaluation.

The data gathered from these experimental studies contribute to understanding levels and routes of exposure, time of administration, multiple target organ toxicity, and combinations of chemicals. The relevance of the toxicity data and extrapolation for human risk assessment, however, requires careful interpretation of the information in order to construct valid conclusions as to health risk. As discussed in later chapters on animal toxicology testing (Part II), organization of animal studies is particularly important in order to extrapolate data for human risk assessment.

4.1.2 DOSE–RESPONSE ASSESSMENT

As the relevant studies are examined for assessment of potential toxicity, the second objective of risk assessment is the determination of the dose-response relationship. Factors used to estimate the dose or doses associated with the appearance of adverse effects are identified according to the determination of the following statistical computations:

> **Quantal relationship** Number of test subjects or individuals in a population that demonstrate an effect; varies with level of exposure (see Figure 2.5, Chapter 2).
> **Graded relationship** Severity of lesions exhibited by the test subjects or individuals; is proportional to dose.
> **Continuous relationship** Alteration of a physiological or biological indicator (e.g., body weight) with respect to dosage regimen.

Accordingly, definitive proof of a dose–response relationship is understood as a statistically positive correlation between the intensity or frequency of the effect as the dose increases. An important factor that influences interpretation of the relationship involves the recognition of experimental thresholds for chemicals in animal studies and demonstration of similar thresholds in human populations. Thresholds define acceptable exposure levels within human populations. Consequently, the selection of adequate numbers of animal species, use of a variety of exposure levels, and screening with valid *in vitro* techniques allow for improved assessment of threshold estimates from experimental data.

4.1.3 EXPOSURE ASSESSMENT

Exposure assessment is an estimation of the intensity, frequency, and duration of human exposure or the potential of exposure to a toxic substance present in the environment. A variety of exposure possibilities that describe the magnitude, duration, schedule, and route of exposure to a substance are determined. Exposure assessment also involves determination of the extent of exposure and the circumstances surrounding the type of release. Available toxicity data, current limits of environmental pollution, and estimations of current or projected intakes are included in the parameters. The information gathered attempts to address the uncertainties associated with the conclusions about exposure assessment in particular and risk assessment in general.

4.1.4 RISK CHARACTERIZATION

Risk characterization involves the accumulation of data for use in estimating the probable incidence of an adverse health effect to humans under various conditions of exposure. This component of risk assessment incorporates the most reliable human databases and adequately reviewed animal studies. Together with hazard identification, dose–response relationship, and exposure assessment, risk characterization enables the formation of model systems used in risk assessment prediction.

4.1.5 THRESHOLD RELATIONSHIPS

Estimation of low dose exposure levels and determining their effects in human populations are particularly difficult processes. Threshold values represent a range of the lowest possible exposures that are tolerated by humans in the environment or in an occupational setting without risk of adverse health effect. The dose–response relationship at low concentrations, however, does not always calculate with good correlations because effects are not necessarily linear at those levels. Consequently, data obtained from chronic *in vivo* and valid *in vitro* studies can support the estimation of threshold values. Table 4.1 summarizes threshold relationships and applications in risk assessment.

TABLE 4.1
Threshold Values and Associated Toxic Classifications

Classification	Type	Description	Applications	Method of Calculation
Threshold	Non-zero threshold	Adverse health effects unlikely below this level	Systemic organ toxicity; germ-cell mutations; developmental and reproductive toxicity	Safety (uncertainty) factor method
Non-threshold	Zero threshold	Adverse health effects possible at any dosage level	Mutagenic and carcinogenic effects	Risk analysis method

TABLE 4.2
U.S. Regulatory Agencies and Acts under Their Jurisdictions

Agency	Act	Regulatory Responsibility
NRC	Energy Reorganization Act of 1974	Act established NRC; development and production of nuclear weapons, promotion of nuclear power, and other energy-related work
EPA	Clean Air Act	Air quality, air pollution
	Clean Water Act	Water pollution, waste treatment management, toxic pollutants
	CERCLA	Established Superfund; clean-up of hazardous substances released in air, land, water
	FIFRA	Safety and regulation of pesticides
	RCRA	Generation, transportation, and disposal of hazardous waste
	Safe Drinking Water Act	Standards for drinking water; establishes MCLs
	Toxic Substances Control Act	Production, processing, importation, testing and use of potentially hazardous chemicals; testing for HVHE and HPV chemicals
FDA	Federal Food, Drug and Cosmetic Act	Safety and/or efficacy of food and color additives, medical devices, pre-marketing drug approval, cosmetics
DEA	Controlled Substances Act	Manufacture, distribution, dispensing, registration, handling of narcotics, stimulants, depressants, hallucinogens, and anabolic steroids; monitors chemicals used in illicit production of controlled substances
CPSC	Consumer Product Safety Act	Safety standards for consumer products
	FHSA	Toxic, corrosive, radioactive, combustible, or pressurized labeling requirements for hazardous substances
	PPPA	Packaging of hazardous household products
OSHA	OSHA	Sets occupational safety and health standards and toxic chemical exposure limits for working conditions

Note: CERCLA = Comprehensive Environmental Response, Compensation and Liability Act. DEA = Drug Enforcement Administration. EPA = Environmental Protection Agency. FDA = Food and Drug Administration. FHSA = Federal Hazardous Substances Act. FIFRA = Federal Insecticide, Fungicide and Rodenticide Act. HPV = high production volume. HVHE = high volume–high exposure. MCLs = maximum contaminant levels for drinking water. NRC = Nuclear Regulatory Commission. OSHA = Occupational Safety and Health Administration (or Act). PPPA = Poison Prevention Packaging Act. RCRA = Resource Conservation and Recovery Act.

4.2   REGULATORY TOXICOLOGY

Several administrative regulatory agencies are responsible for the application, enforcement, and establishment of rules and regulations associated with safe and effective chemical and drug use in the United States. Table 4.2 lists the relevant regulatory agencies and the acts under their jurisdiction.

4.2.1 NUCLEAR REGULATORY COMMISSION (NRC)

Headed by a five-member commission, the NRC is an independent agency established by the Energy Reorganization Act of 1974 to regulate civilian use of nuclear materials.* The NRC's primary mission is to protect public health and safety and the environment from the effects of radiation from nuclear reactors, materials, and waste facilities. In addition, it monitors national defense and promotes security from radiological threat. NRC carries out its mission through the enactment of policy; protection of workers from radiation hazards; development of standards; inspection of facilities; investigation of cases; research and development; licensing the procurement, storage, and use of radiation materials; and adjudication.

4.2.2 ENVIRONMENTAL PROTECTION AGENCY (EPA)

The EPA is headed by a presidentially appointed administrator. Its mission is to protect human health and safeguard the natural environment. The agency employs over 18,000 engineers, scientists, environmental protection specialists, and staff in program offices, regional offices, and laboratories.

EPA's mandate arises from federal laws protecting human health and the environment. It oversees natural resources, human health, economic growth, energy, transportation, agriculture, industry, and international trade in establishing environmental policy. EPA also provides leadership in environmental science, research, education and assessment efforts, and works closely with other federal agencies, state and local governments, and Indian tribes to develop and enforce regulations. EPA sets national standards and delegates responsibility to states and tribes for issuing permits, enforcing compliance, and issuing sanctions. It monitors and enforces a variety of voluntary pollution prevention programs and energy conservation efforts, particularly with industrial concerns.

4.2.3 FOOD AND DRUG ADMINISTRATION (FDA)

At the beginning of the 20th century, revelations about filth in the Chicago stockyards shocked the nation into the awareness that protection against unsafe products is beyond any individual's means in an industrial economy. The U.S. Congress responded to Upton Sinclair's best-selling book titled *The Jungle* by passing the Food and Drug Act of 1906 that prohibited interstate commerce of misbranded and adulterated foods and drugs.

Enforcement of the law was entrusted to the U.S. Department of Agriculture's Bureau of Chemistry, which later became the FDA. The 1906 act was the first of more than 200 laws, some of which are discussed below, that constitute a compre-

---

* The Atomic Energy Act of 1954 established a single agency known as the Atomic Energy Commission that was responsible for the development and production of nuclear weapons and also for the development and safety regulation of civilian uses of nuclear materials. The Act of 1974 split these functions and assigned the responsibility for the development and production of nuclear weapons and other energy-related work to the Department of Energy, while delegating the regulatory work to the NRC.

hensive and effective network of public health and consumer protections. FDA's mission, therefore, is to promote and protect public health by ensuring the safety and efficacy of products in the market and monitoring them for continued safety.

Overall, the FDA regulates $1 trillion worth of products a year. It is under the jurisdiction of the Department of Health and Human Services (HHS) and consists of nine centers: the Center for Biologics Evaluation and Research (CBER), the Center for Devices and Radiological Health (CDRH), the Center for Drug Evaluation and Research (CDER), the Center for Food Safety and Applied Nutrition (CFSAN), the Center for Veterinary Medicine (CVM), the National Center for Toxicological Research (NCTR), the Office of the Commissioner (OC), and the Office of Regulatory Affairs (ORA).

The Code of Federal Regulations (CFR) is a codification of the rules published in the *Federal Register* by executive departments and agencies of the U.S. government. The CFR is divided into 50 titles (volumes) that represent broad areas subject to federal regulation, with environmental regulations contained mainly in Title 40. Products regulated by the FDA through the CFR include food and food additives (except for meat, poultry, and certain egg products); medical and surgical devices; therapeutic drugs; biological products (including blood, vaccines, and tissues for transplantation); animal drugs and feed; and radiation-emitting consumer and medical products. The FDA also acts to prevent the willful contamination of all regulated products and improve the availability of medications to prevent or treat injuries caused by biological, chemical, and nuclear agents.

The federal Food, Drug, and Cosmetic Act (FD&C) of 1938 was passed after a legally marketed toxic elixir killed 107 children and adults. The FD&C Act authorized the FDA to demand evidence of safety for new drugs, issue standards for foods, and conduct factory inspections.

The Kefauver-Harris Amendments of 1962 were inspired by the thalidomide tragedy in Europe. The amendments strengthened the rules for drug safety and required manufacturers to prove drug effectiveness.*

The Medical Device Amendments of 1976 followed a U.S. Senate finding that faulty medical devices had caused 10,000 injuries, including 731 deaths. These amendments established guidelines for safety and effectiveness of new medical devices.

4.2.4 DRUG ENFORCEMENT ADMINISTRATION (DEA)

The DEA is under the policy guidance of the Secretary of State and HHS and is headed by an administrator. Its mission is to enforce controlled substances laws and regulations and prosecute individuals and organizations involved in the growing, manufacture, or distribution of controlled substances appearing in or destined for illicit traffic. It also recommends and supports non-enforcement programs aimed at reducing the availability of illicit controlled substances in domestic and international markets. The DEA's primary responsibilities include:

---

* Prompted by the FDA's vigilance and monitoring of clinical cases, thalidomide was prevented from entering the U.S. market.

1. Investigation and preparation for the prosecution of major violators of controlled substance laws operating at interstate and international levels
2. Prosecution of criminals and drug gangs
3. Management of a national drug intelligence program in cooperation with federal, state, local, and foreign officials to collect, analyze, and disseminate strategic and operational drug intelligence information
4. Seizure of assets derived from, traceable to, or intended to be used for illicit drug trafficking
5. Enforcement of provisions of the Controlled Substances Act (see below) as they pertain to the manufacture, distribution, and dispensing of legally produced controlled substances
6. Coordination and cooperation with federal, state, and local law enforcement officials on mutual drug enforcement efforts including the reduction of the availability of illicit abuse-type drugs, crop eradication, and training of foreign officials

The Controlled Substances Act (CSA), Title II of the Comprehensive Drug Abuse Prevention and Control Act of 1970, is a consolidation of numerous laws. It regulates the manufacture and distribution of narcotics, stimulants, depressants, hallucinogens, anabolic steroids, and chemicals used in the illicit sale or production of controlled substances. The CSA places all regulated substances onto one of five schedules, based upon therapeutic importance, toxicity, and potential for abuse or addiction. Table 4.3 describes the schedules and lists examples of drugs within these classifications. Regulatory proceedings initiated by the DEA or HHS may add, delete, or change the schedule of a drug or other substance. Other interested parties including drug manufacturers, medical and pharmacy societies or associations, public interest groups concerned with drug abuse, state and local government agencies, and individuals may also petition the DEA to modify a schedule.

The CSA also creates a closed system of distribution for those authorized to handle controlled substances. The cornerstone of this system is the registration of all those authorized by the DEA to handle such substances. All registered individuals and firms are required to maintain complete and accurate inventories and records of all transactions involving controlled substances and provide security for the storage of controlled substances.

4.2.5 CONSUMER PRODUCTS SAFETY COMMISSION (CPSC)

The CPSC is an independent federal regulatory agency created by Congress in 1972 under the Consumer Product Safety Act.* Its mission is to protect the public from unreasonable risks of injuries and deaths associated with some 15,000 types of consumer products by informing the public about product hazards through local and

* The CPSC is headed by three presidential nominees who serve as commissioners (one is the chair person). Commissioners are confirmed by the Senate for staggered 7-year terms. The commissioners set policy. The CPSC is not part of any other governmental department or agency. The Congressional Affairs, Equal Employment and Minority Enterprise, General Counsel, Inspector General, Secretary, and Executive Director groups report directly to the chairman.

TABLE 4.3
Controlled Substance Schedules and Descriptions

Schedule	Description of Regulated Substances	Examples of Listed Substances
I	Drug or substance has high potential for abuse; has no currently accepted medical use in treatment in U.S.; lack of accepted safety for use under medical supervision	Benzylmorphine, etorphine, heroin, dimethyltryptamine (DMT), marijuana, lysergic acid diethylamide (LSD), mescaline, peyote, psilocybin, cocaine, tetrahydrocannabinols (THCs)
II	Drug or substance has high potential for abuse; has currently accepted medical use in treatment in U.S. with or without severe restrictions; may lead to severe psychological or physical dependence	Fentanyl, levorphanol. methadone, opium and derivatives
III	Drug or substance has potential for abuse less than substances in schedules I and II; has currently accepted medical use in treatment in U.S.; may lead to moderate or low physical dependence or high psychological dependence	Amphetamine, phenmetrazine methylphenidate, phencyclidine, nalorphine, anabolic steroids
IV	Drug or substance has low potential for abuse relative to drugs or substances in schedule III; has currently accepted medical use in treatment in U.S.; may lead to limited physical or psychological dependence	Barbital, chloral hydrate, meprobamate, phenobarbital
V	Drug or substance has low potential for abuse; has currently accepted medical use in treatment in U.S.; may lead to limited physical or psychological dependence	Not >200 mg codeine/100 ml; not >2.5 mg diphenoxylate and not <25 µg/dosage unit atropine sulfate (Lomotil®)

national media coverage, publication of booklets and product alerts, a web site, a telephone hotline, the National Injury Information Clearinghouse, its Public Information Center, and responses to the Freedom of Information Act (FOIA) requests.

The CPSC fulfills its mission through the development of voluntary standards with industry and by issuing and enforcing mandatory standards or banning consumer products if no feasible standard adequately protects the public. The CPSC obtains recalls of products or arranges for product modifications. It also conducts research on potential product hazards, informs and educates consumers, and responds to consumer inquiries. Table 4.4 lists some of the product guidelines under CPSC jurisdiction.

CPSC administers the Poison Prevention Packaging Act (PPPA) of 1970. The PPPA requires child-resistant packaging of hazardous household products. The number of reported deaths from ingestions of toxic household substances by children since the inception of this act has declined significantly. However, it is estimated that more than 1 million calls to poison control centers are still registered following unintentional exposure to medicines and household chemicals by children under 5

TABLE 4.4
Consumer Products and Established Safety and Monitoring Guidelines under CPSC Jurisdiction

All-terrain vehicles	Home heating equipment
Arts and crafts	Household products
Bicycles	Indoor air quality
Child safety	Older consumer safety
Children's furniture	Outdoor power equipment
Clothing	Playground safety
Consumer product safety reviews	Poison prevention
Crib safety and SIDS reduction	Pool and spa safety
Electrical safety	Public use products
Fire safety	Recreational and sports safety
General information	Reports
Holiday safety	Toys

Note: SIDS = sudden infant death syndrome.

---

years of age. More than 85,000 children are examined in emergency departments and almost 50 child deaths result from this type of exposure each year.*

Some of the reasons accounting for continuing ingestions include the availability of non-child-resistant packaging on request for prescription medications and generally for over-the-counter medications; inadequate quality control by manufacturers leading to defective child-resistant closures; misuse of child-resistant packaging in the home (failure to replace or secure caps, transferring contents to non-child-resistant packages); and violations by health professionals. Consequently, the CPSC has designed a textbook to educate health professionals, particularly pharmacists and physicians, about the child-resistant packaging program. It is intended to be incorporated into medical and pharmacy school curricula to bring greater awareness of their legal responsibilities. Table 4.5 lists some of the substances covered by the PPPA regulations.

4.2.6 OCCUPATIONAL SAFETY AND HEALTH ADMINISTRATION (OSHA)

Created by the Occupational Safety and Health Act of 1970, OSHA assures safe and healthful working conditions. It accomplishes this task through inspectors and staff personnel. OSHA authorizes the enforcement of the standards developed under the act in cooperation with 26 states. In addition, it provides for research, information, education, and training in occupational safety and health. As a testament to the impact of the agency and the act, workplace fatalities and occupational injury and

---

* According to the National Electronic Injury Surveillance System (a CPSC database tracking emergency room visits) and the American Association of Poison Control Centers (2003).

TABLE 4.5
Drugs and Substances Covered by PPPA of 1970

Therapeutic Category	Examples of Drugs and Substances
Analgesics	Aspirin, acetaminophen, methyl salicylate
Controlled substances	Opioids, S/H, stimulants
Vitamins and dietary supplements	Iron-containing drugs and other dietary supplements
Local anesthetics	Lidocaine, dibucaine, and minoxidil
Non-prescription NSAIDs	Ibuprofen, naproxen, and ketoprofen
Non-prescription antihistamines	Diphenhydramine (Benadryl®)
Non-prescription anti-diarrhea products	Loperamide (Imodium®)
General	Human oral prescription drugs (with some exceptions and exemptions)

Note: NSAIDs = non-steroidal anti-inflammatory drugs, S/H = sedative/hypnotics.

illness rates have been reduced by 40 to 50% while U.S. employment doubled from 56 to 111 million workers at 7 million sites since 1971.

The agency fulfills its mandate by setting standards established by regulations. For example, the OSHA Lead Standards for General Industry and Construction require employers to provide biological monitoring for workers exposed to airborne lead above the action level. Employers must provide monitoring of lead and zinc protoporphyrin (or free erythrocyte protoporphyrin) levels in blood. The analyses must be performed by laboratories that meet accuracy requirements specified by OSHA. The OSHA List of Laboratories Approved for Blood Lead Analysis is designed to serve as a source to locate laboratories that OSHA has determined meet the requirements of the accuracy provisions of the lead standards. Laboratories voluntarily provide proficiency test data to OSHA for evaluation.

SUGGESTED READINGS

Clarkson, T.W., Principles of risk assessment, *Adv. Dent. Res.,* 6, 22, 1992.
Collins, T.F., History and evolution of reproductive and developmental toxicology guidelines, *Curr. Pharm. Des.,* 12, 1449, 2006.
Cross, J. et al., Postmarketing drug dosage changes of 499 FDA-approved new molecular entities, *Pharmacoepidemiol. Drug Saf.,* 11, 439, 2002.
Dietz, F.K., Ramsey, J.C., and Watanabe, P.G., Relevance of experimental studies to human risk, *Environ. Health Perspect.,* 52, 9, 1983.
Hays, S.M., Becker, R.A., Leung, H.W., Aylward, L.L., and Pyatt, D.W., Biomonitoring equivalents: a screening approach for interpreting biomonitoring results from a public health risk perspective, *Regul. Toxicol. Pharmacol.,* 2006 [Epub ahead of print].
Luft, J. and Bode, G., Integration of safety pharmacology endpoints into toxicology studies, *Fundam. Clin. Pharmacol.,* 16, 91, 2002.
Nicoll, A. and Murray, V., Health protection: strategy and a national agency, *Publ.Health,* 116, 129, 2002.

Patterson, J., Hakkinen, P.J., and Wullenweber, A.E., Human health risk assessment: selected
Internet and world wide web resources, *Toxicology*, 173, 123, 2002.

REVIEW ARTICLES

Acquavella, J. et al., Epidemiologic studies of occupational pesticide exposure and cancer:
regulatory risk assessments and biologic plausibility, *Ann. Epidemiol.*, 13, 1, 2003.
Baht, R.V. and Moy, G.G., Monitoring and assessment of dietary exposure to chemical
contaminants, *World Health Stat. Q.*, 50, 132, 1997.
Barlow, S.M. et al., Hazard identification by methods of animal-based toxicology, *Food Chem.
Toxicol.*, 40, 145, 2002.
Bernauer, U., Oberemm, A., Madle, S., and Gundert-Remy, U., The use of *in vitro* data in
risk assessment, *Basic Clin. Pharmacol. Toxicol.*, 96, 176, 2005.
Boyes, W.K. et al., EPA's neurotoxicity risk assessment guidelines, *Fundam. Appl. Toxicol.*,
40, 175, 1997.
Buchanan, J.R., Burka, L.T., and Melnick, R.L., Purpose and guidelines for toxicokinetic studies
within the National Toxicology Program, *Environ. Health Perspect.*, 105, 468, 1997.
Bull, R.J., Conolly, R.B., De Marini, D.M., MacPhail, R.C., Ohanian, E.V., and Swenberg,
J.A. Incorporating biologically based models into assessments of risk from chemical
contaminants, *J. Am. Water Works Assn.*, 85, 49, 1993.
Carere, A. and Benigni, R., Strategies and governmental regulations, *Teratog. Carcinog.
Mutagen,* 10, 199, 1990.
Casciano, D.A., FDA: a science-based agency, *FDA Consum.*, 36, 40, 2002.
Cimino, M.C., Comparative overview of current international strategies and guidelines for genetic
toxicology testing for regulatory purposes, *Environ. Mol. Mutagen.* 47, 362, 2006.
Clewell, H.J. 3rd, Andersen, M.E., and Barton, H.A., A consistent approach for the application
of pharmacokinetic modeling in cancer and noncancer risk assessment, *Environ.
Health Perspect.*, 110, 85, 2002.
Conolly, R.B., The use of biologically based modeling in risk assessment, *Toxicology*, 181,
275, 2002.
Cornfield, J., Rai, K., and Van Ryzin, J., Procedures for assessing risk at low levels of exposure,
Arch. Toxicol. Suppl. 3, 295, 1980.
Dorato, M.A. and Engelhardt, J.A., The no-observed-adverse-effect level in drug safety
evaluations: use, issues, and definition(s), *Regul. Toxicol. Pharmacol.* 42, 265, 2005.
Eason, C. and O'Halloran, K., Biomarkers in toxicology versus ecological risk assessment,
Toxicology, 181, 517, 2002.
Indans, I., The use and interpretation of *in vitro* data in regulatory toxicology: cosmetics,
toiletries and household products, *Toxicol. Lett.*, 127, 177, 2002.
Jensen, J. and Pedersen, M.B. Ecological risk assessment of contaminated soil, *Rev. Environ.
Contam. Toxicol.*, 186, 73, 2006.
Kowalski, L. et al., Overview of EPA Superfund human health research program, *Int. J. Hyg.
Environ. Health,* 205, 143, 2002.
Lathers, C.M., Risk assessment in regulatory policy making for human and veterinary public
health, *J. Clin. Pharmacol.*, 42, 846, 2002.
Hansson, S.O. and Ruden, C., Evaluating the risk decision process, *Toxicology*, 218, 100, 2006.
Hayashi, Y., Designing *in vitro* assay systems for hazard characterization: basic strategies and
related technical issues, *Exp. Toxicol. Pathol.*, 57, 227, 2005.

Liebsch, M. and Spielmann, H., Currently available *in vitro* methods used in the regulatory toxicology, *Toxicol. Lett.,* 127, 127, 2002.

Lorber, M., Indirect exposure assessment at the United States Environmental Protection Agency, *Toxicol. Ind. Health,* 17, 145, 2001.

Lu, F.C. and Sielken, R.L. Jr., Assessment of safety/risk of chemicals: inception and evolution of the ADI and dose-response modeling procedures, *Toxicol. Lett.,* 59, 5, 1991.

Pirkle, J.L. et al., Using biological monitoring to assess human exposure to priority toxicants, *Environ. Health Perspect.,* 103, 45, 1995.

Temple, R., Policy developments in regulatory approval, *Stat. Med.,* 21, 2939, 2002.

Van Ryzin, J., Quantitative risk assessment, *J. Occup. Med.,* 22, 321, 1980.

Weed, D.L., Weight of evidence: a review of concept and methods, *Risk Anal.,* 25, 1545, 2005.

5 Descriptive Animal Toxicology Tests

5.1 CORRELATION WITH HUMAN EXPOSURE

5.1.1 HUMAN RISK ASSESSMENT

The information derived from descriptive animal toxicology tests is useful in determining the potential toxicity of a compound to humans. The objective of these tests is to identify a toxic chemical at an early stage of chemical development, especially if the substance is already commercially available. Together with *in vitro* tests, animal tests are applied as biological markers of chemically induced risks, whether synthetic or naturally occurring.

These validated batteries of tests are used by regulatory agencies to screen for or predict human toxic effects in an attempt to establish a significant frame of reference for monitoring environmental chemical threats, therapeutic adverse reactions, and commercial and occupational toxicity. Detailed discussions of risk assessment and its application for estimating toxic potential for humans are presented in Chapter 4.

5.1.2 PREDICTIVE TOXICOLOGY AND EXTRAPOLATION TO HUMAN TOXICITY

The ability to predict toxicity in humans and provide a responsible level of safety is probably the most trying conclusion arising from animal testing. Correlation of results of toxicology testing described in animals with human exposure requires careful consideration of the parameters of the animal tests, including the selection of species with similar physiologies. Consideration of animal housing, cost effectiveness, and selection of laboratory diet also calculates into the toxicology testing equation. The variety of tests available to achieve valid conclusions with a particular chemical must be systematically chosen according to the guidelines of the regulatory agencies.

5.2 ANIMAL WELFARE AND U.S. ANIMAL WELFARE ACT

The U.S. Animal Welfare Act (originally enacted in 1966) is promulgated under the Code of Federal Regulations (Title 7, Sections 2131 to 2156). In general, the act authorizes the Secretary of Agriculture to regulate transport, sale, and handling of dogs, cats, nonhuman primates, guinea pigs, hamsters, and rabbits intended to be used in research or "for other purposes." Several amendments have been enacted, including Public Law 91-579 (1970) that expanded the list of animals covered by the act to include all warm-blooded animals used or intended for use in experimentation or exhibition.

Research facilities have been defined and regulations have been developed regarding record keeping and humane care and treatment of animals in or during commerce, exhibition, experimentation, and transport. The regulations also mention inspections and appropriate use of anesthetics, analgesics, and tranquilizers. A further amendment to the act (Public Law 99-198, *Food Security Act of 1985, Subtitle F: Animal Welfare*) is referred to as the Improved Standards for Laboratory Animals Act.

The amendment clarifies what is meant by "humane care" and details specifics such as sanitation, housing, and ventilation. It directs the Secretary of Agriculture to establish regulations to provide exercise for dogs and mandates provisions for establishing an adequate physical environment to promote the psychological health of nonhuman primates. It specifies that pain and distress must be minimized in experimental procedures and requires that principal investigators consider alternatives to such procedures. It also defines experimental practices that are considered painful.

The 1985 act also establishes a requirement for formation of and describes the role and composition of Institutional Animal Care and Use Committees (IACUCs) and defines the responsibilities of the Animal and Plant Health Inspection Service. Also included is the formation of an information service at the National Agricultural Library to assist those regulated by the act in preventing unintended duplication of research, providing employee training, searching for ways to reduce or replace animal testing, and obtaining information on how to decrease pain and distress.

The current version of the regulations, as developed by the U.S. Department of Agriculture, specifies how to comply with the Animal Welfare Act and its amendments. It is divided into four sections: Definitions, Regulations, Standards, and Rules of Practice Governing Proceedings under the Animal Welfare Act. The Definitions section describes the terms used in the legislation. For example, the *animal* term specifically excludes rats of the genus *Rattus* and mice of the genus *Mus* used in research. The Regulations section includes subparts pertaining to licensing, registration, inspection of research facilities, responsibilities of attending veterinarians, and regulations governing adequate veterinary care. It also promulgates guidelines associated with stolen animals, records, compliance with standards and holding periods, and miscellaneous topics such as confiscation and destruction of animals, and access to and inspection of records and property.

The bulk of the third section provides standards for specific species. Included are sections covering cats and dogs, guinea pigs, hamsters, rabbits, nonhuman primates, marine mammals, and the general category of other warm-blooded animals. Standards include those for facilities and operations, health and husbandry systems, and transportation. The final section sets forth the rules of practice applicable to adjudicating administrative proceedings under Section 19 of the Animal Welfare Act (see Suggested Readings and Review Articles listed at the end of this chapter).

5.3 CHEMICALS

5.3.1 SELECTION OF CHEMICALS

The selection of appropriate chemicals for investigation and the aptness of the application of these chemicals for fulfilling the objectives of a study establishes the foundation for an entire project. Commencing the project thus requires a thorough knowledge of the chemicals under consideration including but not limited to the physicochemical properties of the test agents, the suspected or known mechanisms of toxicity, and the applications and uses of the agents in the public sector.

5.3.1.1 Solubility

Physicochemical properties involve knowledge of the physical states of chemicals. This aspect is important not only to demonstrate the commercial availability of a chemical, but also to properly manipulate the agent in a laboratory. For instance, the aqueous or lipid solubility of a chemical determines the choice of a suitable vehicle. Thus, a substance with greater water solubility is more easily dissolved in aqueous media including buffers, electrolytes, or drinking water.

Poor water-soluble chemicals require additional measures in order to deliver the appropriate concentrations to test subjects. Poor water solubility would preclude their inclusion in water vehicles because combination in an aqueous solvent would lead to inaccurate vehicle concentrations. Thus, water-insoluble chemicals are difficult to administer parenterally or in drinking water.

Lipid-soluble chemicals require dissolution in organic solvents either as stock solutions or for whole vehicle delivery. Dissolving a test agent in an organic solvent as a stock solution renders the agent temporarily soluble, after which the extent of solubility in the subsequent water vehicle for *in vivo* administration is effectively determined. The selected organic vehicle is necessarily inert and becomes part of the vehicle control group. Alternatively, insoluble agents in aqueous media necessitate preparing an emulsion or suspension, thus requiring additional manipulations such as shaking, separate storage conditions, observance of shelf life, or homogenization. In any event, whether an agent is dissolved in an organic vehicle or requires formation of an emulsion or suspension, the nature of the preparation should be acceptable for administration via an oral, parenteral, or inhalation route.

5.3.1.2 Vehicle

Limited or poor aqueous solubility necessitates formulation of an emulsion, suspension, or solution in vegetable oils or organic solvents so as to form a homogenous preparation and deliver a correct dose. The use of vegetable oils such as corn, olive, peanut, sunflower, and soy bean as delivery vehicles is not without precautions. Oils generally enhance absorption of orally administered agents more than equivalent concentrations delivered in aqueous media. Thus, chemicals orally administered in oils risk greater target organ toxicity by virtue of experimentally enhanced absorption. In addition, as both vehicle and test substance are metabolized *in vivo*, some influence on chemical interactions is possible. Several studies with pesticides, herbicides, polyethylene glycols, and alcohols revealed unanticipated results when administered orally or parenterally in oil-based vehicles (see Suggested Readings and Review Articles at the end of this chapter).

Similarly, chemicals dissolved in oil-based vehicles are well suited for dermal applications. Preparation of solutions, suspensions, or emulsions in oil or organic vehicles ensures more efficient penetration of the skin and enhanced dermal absorption. This technique of dermal administration is limited by the method of application to the animal. Several models developed for dermal testing in animals secure proper dermal delivery of test agents.

As with gastrointestinal or parenteral administration, proper dissolution in an appropriate vehicle is necessary for nasal or respiratory inhalation of toxicants. The vehicle and dissolved agent are administered through any of several methods appropriate for inhalation studies including exposure of animals to high pressure aerosolization or release of vapors in contained, enclosed chambers. Consequently, with agents that have poor water solubility, the interaction of the vehicle with pulmonary tissue may influence mucous membrane absorption. For instance, oil-based vehicles cause local upper respiratory irritation or inflammation of lower pulmonary airways, resulting in respiratory inflammatory reactions. Inflammation unrelated to the test agent thus must be distinguished from reactions caused by the chemical. In addition, insoluble solid particles present in a suspension or emulsion confound interpretation of results and require additional testing. A degradation product may interact with a test chemical or may be erroneously considered as a metabolite formed *in vivo*.

5.3.2 ROUTE OF ADMINISTRATION

Selection of the appropriate route of administration is important in toxicological analysis and evaluation of the degree of toxicity. In general, as the objective of a toxicology study attempts to mimic human or animal risk exposure, the route of administration of a chemical to a test species should correlate with normal circumstances encountered for human exposure. For example, studies that attempt to understand toxicant exposure for human occupational toxicology require development of toxicity testing protocols that offer oral ingestion, dermal contact, and inhalation as important routes of possible contact.

Alternatively, domestic environmental exposure studies involving pesticides, herbicides, or fungicides should use an inhalation or dermal exposure animal model. Hence, development of animal models for toxicology testing should consider incorporating a parallel route of administration that imitates human exposure.

5.4 SPECIES DIFFERENTIATION

5.4.1 SELECTION OF APPROPRIATE ANIMAL SPECIES

As with the selection of an appropriate route of administration in designing toxicological studies, it is important to consider several criteria before selecting a suitable animal species. The criteria include:

1. Classification of the toxic agent according to human exposure (see Part I, Chapter 2, Effects of Chemicals)

2. Anatomic, physiologic, and metabolic similarity of the animal species to humans
3. Ages, life expectancies, and sexes of the animals
4. Objectives of the experiment
5. Previous experience with the species, convenience in handling the size of the animal, housing requirements, and daily care for the species
6. Cost efficiency of toxicology study

It should be emphasized that animal models are selected based on their ability to substitute for the effects seen with chemical exposure to humans. The classification of the chemical determines the route of administration, duration, and dosage and subsequently influences species selection. For acute toxicity studies, the rodent model is most appropriate based on versatility of administration and the short duration corresponding to the physiological parameters of rodents. Dosage in any animal species must be adjusted based on metabolic rate and historical dosages required for rodent species.

Although no perfect surrogate for humans exists, each species contributes distinct features that underscore biological and physiological differences among the various groups. It is generally accepted that differences in absorption, distribution, biotransformation, and elimination of chemicals exist among animal species and from humans, thus necessitating careful interpretation of toxicological data. Table 5.1 lists some of the major characteristics of mammalian animal species commonly used in toxicological investigations and considerations affecting the selection and incorporation of animals in toxicology testing studies.

It is common knowledge that no single species accurately reflects the complex characteristics of human physiology and metabolic activity. Inter- and intraspecies variability is expected. In fact, this variability exists even within the same species, especially when the parameters of a study have not been adequately controlled. Careful attention to criteria used in repeating studies (time of day when animals are handled, sexes, and numbers of animals used) will minimize species variation. The goal of using animal models in toxicological assessment, therefore, is to produce results that can be extrapolated to humans and reflect the most significant correlations with the objectives of a study. As will be shown in later chapters, if the objective of a study is demonstration of risk for a suspected carcinogen in humans, the interpretation of results is based on the understanding of the differences between the animal models and humans.

Interspecies variability is also minimized by selecting animals bred under controlled environments. Selection of animals with the following uniform characteristics aids in decreasing experimental variability: (1) use of inbred (homogeneous) or outbred (heterogeneous) strains of the animal*; (2) uniform mean body weights; (3)

---

* Outbred strains of rodents are closed colonies propagated through random mating in an established breeding system, avoiding the mating of close relatives. While a colony of animals is reasonably uniform in characteristics, each individual is genetically distinct. Outbred strains show genetic variabilities among colonies from different sources even though they may bear a generic name (Wistar, Sprague-Dawley, or Swiss albino). They vary phenotypically and, even within a single colony, genetic characteristics change with time and breeding within colonies.

Principles of Toxicology Testing

TABLE 5.1
Biological and Physiological Differences among Mammalian Species Commonly Used in Toxicological Investigations and Humans

Characteristics[a]	Species				
	Mouse	Rat	Rabbit	Dog	Human
Adult weight	18 to 40 g	250 to 800 g	2.5 to 5 kg	Varies	75 kg
Life span (yr)	1 to 3	2 to 3.5	5 to 10	16 to 18	78
Number of chromosomes	40	42	44	78	46
Body temperature (°C)	37.1 to 37.4	37.1 to 37.4	38 to 40	38 to 40	37
Age at puberty	18 to 49 days	20 to 50 days	8 to 22 wk	8 to 14 mo	12 to 15 yr
Gestation (days)	19 to 21	21 to 22	31 to 32	63	270
Litter size	4 to 12	6 to 14	4 to 10	2 to 12	~1
Heart rate (bpm)	310 to 840	320 to 480	150 to 300	70 to 130	60 to 90
Systolic/diastolic blood pressure (mmHg)	145/105	100/75	110/80	145/82	120/80
Whole blood volume (ml/100 g)	5.8	5.6 to 7.1	6.0	6.0 to 7.0	6.5 to 7.5
Respiratory rate (rpm)	163	85 to 110	32 to 65	10 to 30	12 to 18
Plasma pH	7.2–7.4	7.4	7.4	7.4	7.4
Other characteristics[b]	Barbering, water deprivation, fighting, weight loss	Similar to mouse; lacks gall bladder, appendix; poor vision	Crepuscular activity;[c] anti-serum production, pyrogen testing, cardiovascular studies	Model for various human diseases (cardiovascular, diabetes, behavioral)	Anatomically similar to rodent

[a] Average values for male and female.
[b] In group housed animals (see text).
[c] Crepuscular = most active at dawn, twilight.

dependable yearly survival rates; (4) similar incidence of age-related diseases; (5) similar occurrence of spontaneous tumor formation; and (6) responses to standard diet. Knowledge and assurances by the breeders that these parameters are strictly monitored will decrease unwarranted interspecies variability.

5.4.2 COST EFFECTIVENESS

One of the most important criteria for selecting an appropriate animal species is cost. Animal testing is more expensive than *in vitro* methods. A budget for animal testing must account for:

1. Procurement of animals, housing, and maintenance within the animal care facility
2. Daily requirements for food, water, bedding
3. Employment and training of animal care staff
4. On-staff or on-call veterinarian
5. Adherence to proper procedures for removal and disposal of specimens and waste

The decision to institute, maintain, and develop animal toxicology studies will involve a variety of expected and unexpected costs. These expenses are outlined in Table 5.2, and include the costs of purchasing and shipping, providing and maintaining housing accommodations, animal feed, and supplies, technical care, and utilities. The overhead associated with establishing and maintaining an animal care facility may be prohibitive, especially for smaller facilities. Interestingly, the budgets required for animal toxicology testing have, in part, prompted the development of *in vitro* alternative methods with comparatively less expensive requirements.

Accordingly, based on the predicted costs of acute and chronic studies, ease of handling, convenient housing requirements, and the vast comparative databases available for most chemicals tested using laboratory rodents, they are the most commonly used models for toxicity testing. The decision for using rodents is based more on cost than on their biological similarities to humans. Based on these criteria, rodents are generally appropriate models for toxicology testing. However, it is

TABLE 5.2
Cost-Effective Considerations for Animal Toxicity Studies

Consideration	Description
Animals	Procurement, housing, maintenance
Daily animal requirements	Food, water, bedding
Housing expenses	Utilities, animal care
Technical staff	Technical and auxiliary care, training, employment expenses
Medical care	Veterinarian services
Waste disposal	Contract company for disposal of specimens and waste
IACUC	Indirect and direct costs associated with viable committee

important to note that cost is not an adequate substitute for extensive experience with the biological parameters of animal models, target organ toxicity, classification of chemicals, and toxicokinetics. Selection of rodents as test models, therefore, must include a comparative analysis of the advantages and disadvantages of other animal species.

The number of animals used in a test project necessarily contributes to the cost of a project. According to recent survey data (U.S. Public Health Service, 2006) more than 50,000,000 animals were used in toxicology testing studies in the United States in 2005. While this conglomerate number represents an unwieldy statistic, the decision to incorporate a specific number of animals per treatment group depends on the factors mentioned above. Among those factors not listed previously is the probability of occurrence of a toxic insult.

The probability of a toxic effect occurring in a human is inversely proportional to the number of animals required to demonstrate the effect. That is, as the probability of an occurrence in humans decreases, the greater the number of test animals required to demonstrate a parallel risk of occurrence. For human exposure to most environmental pollutants and commercially available chemicals, the probability of occurrence of toxic effects is less than 0.01%. Consequently, this 1:10,000 rate requires several thousands of animals to demonstrate a comparable causal relationship. Since this number of animals in practice is prohibitive, 10 to 25 animals per treatment group (minimum 5 dosage groups per chemical plus controls) are generally used in acute or chronic toxicity studies, respectively. Understandably, the risk of not detecting all the potential adverse effects (false negatives) as a consequence of toxic insult within the dosage ranges used increases with fewer animals.

5.4.3 ANIMAL HOUSING

As noted above, animal housing rules and regulations and other animal care requirements are specified in the U.S. Animal Welfare Act contained in the Code of Federal Regulations (as amended, 1990). The act outlines the conditions required for animal housing in approved animal care facilities. Among the regulations that apply to housing requirements, the rules dictate the physical environment, lighting, number of animals housed per square foot based on animal and cage size, and job specifications for animal care personnel.

5.4.4 DIET

It has been consistently shown that long-term calorie restriction (CR) is the only intervention that slows the rate of aging and increases mean and maximum life spans in short-lived species. CR extends life spans and retards age-related chronic diseases in a variety of species including rats, mice, fish, flies, worms, and yeasts. Although the mechanism for this occurrence is unclear, it is well understood that CR reduces metabolic rates and oxidative stress, improves insulin sensitivity, and alters neuroendocrine and sympathetic nervous system functions. Rodent studies demonstrate that CR opposes the development of many age-associated pathophysiological changes including detrimental alterations in brain function, learning, and behavior. Long-term CR in monkeys has shown beneficial effects similar to those seen in rodents. Implications associated with the benefits of long-term CR in longer-lived species such as primates and humans are materializing.

It is now generally accepted that rodents fed 8 to 20% less than the *ad libitum* intake exhibit weight loss, have increased longevity, and are less susceptible to a variety of diseases and age-related changes in organ function. In addition, decreased incidence of spontaneous and chemically induced tumor formation appears to correlate with reduction in total caloric content, rather than the elimination of any single micronutrient or macronutrient. Thus, the conditions under which laboratory animals are maintained can significantly influence the results of toxicology studies used for risk assessment.

Nutrition is of importance in toxicological bioassays and research because diet composition and the conditions of administration affect the metabolism and activities of xenobiotic test substances and alter the results and reproducibility of long-term studies. *Ad libitum*-fed animals are not well controlled subjects for experimental studies and this type of feeding leads to considerable inter-laboratory variability. In fact, the U.S. Food and Drug Administration recently addressed the problems associated with uncontrolled food consumption and suggested the levels of dietary control appropriate to achieve standardized growth curves.

The evidence suggests that the primary adaptation appears to be a rhythmic hypercorticism in the absence of elevated ACTH levels. This characteristic hypercorticism evokes a spectrum of responses including (1) decreased glucose uptake and metabolism by peripheral tissues, (2) decreased mitogenic response, (3) reduced inflammatory response, (4) reduced oxidative damage to proteins and DNA, (5) reduced reproductive capacity, and (6) altered drug metabolizing enzyme expression.

Mechanistic data from dietary restriction studies suggest that obese animals are more susceptible to chemically induced toxicity, thus presenting problems in comparability of information. This is especially important in comparing data generated today from a historical databases. Recent studies testing this paradigm demonstrated that dietary control increases animal survival in 2 year studies and also increases bioassay sensitivity. Consequently, devising experimental toxicology protocols that may be influenced by dietary components, especially in young animals, has a direct action on a variety of metabolic factors.

Metabolically, CR reduces the hyperinsulinemia characteristically seen with *ad libitum*-fed rats, subsequently decreasing the regulatory expression of metabolizing enzymes. CR markedly decreases the acute toxicities of several agents, the mechanism of which is mediated through enhanced immunosuppression, reduced rate of metabolism of test chemicals, or prevention of biotransformation. In addition, suspected or known mammalian carcinogens show significantly lower incidence of cancer induction in CR animals. CR also interferes with normal circadian rhythms, as rodents appear to consume their allotted food rapidly when on reduced diets. Interestingly, CR and changes in feeding behavior alter the circadian cycles of serum hormones such as insulin, thyroid hormone, and corticosterone. Overall, CR suppresses xenobiotic metabolism in young animals and exerts a moderate effect in maintaining these same enzymes in older animals.

5.4.5 INSTITUTIONAL ANIMAL CARE AND USE COMMITTEES (IACUCs)

The U.S. Public Health Service (PHS) Policy on Humane Care and Use of Laboratory Animals was promulgated in 1986 and implemented the Health Research Extension Act of 1985 (Public Law 99-158, titled Animals in Research). The Office of Laboratory Animal Welfare (OLAW) at the U.S. National Institutes of Health (NIH), which has responsibility for the general administration and coordination of policy on behalf of the PHS, provides specific guidance, instruction, and materials to institutions on the utilization and care of vertebrate animals used in testing, research, and training. The PHS policy requires institutions to establish and maintain proper measures to ensure the appropriate care and incorporation of all animals involved in research, research training, and biological testing activities conducted or supported by the PHS.

In addition, any activity involving animals must include a written assurance acceptable to the PHS setting forth compliance with this policy. Assurances are submitted to OLAW at the NIH, and are evaluated by OLAW to determine the adequacy of an institution's proposed program for the care and use of animals in PHS-conducted or -supported activities. In summary, the program assurance must include the following information:

1. A list of all branches and major components of the institution.
2. Lines of authority and responsibilities for administering the program.
3. The qualifications, authorities, and responsibilities of veterinarians who participate in the program.
4. The membership list of the IACUC established in accordance with the requirements.
5. The procedures that the IACUC follows to fulfill the requirements.
6. The health program for personnel who work in laboratory animal facilities or have frequent contacts with animals.
7. A synopsis of the training or instruction in the humane practice of animal care and use. Training or instruction in research or testing methods that minimize the number of animals required to obtain valid results and minimize animal distress must be offered to scientists, animal technicians, and other personnel involved in animal care, treatment, or use.
8. The gross square footage of each animal facility, the species housed, an average daily inventory of animals in each facility, and any other pertinent information requested by OLAW.

Other functions of the IACUC include periodic reviews of the institution's program, inspection of facilities, preparation of reports, and submittal of suggestions and recommendations on improving the program.

5.5 METHODOLOGIES

5.5.1 ROUTES OF EXPOSURE, DURATION, AND FREQUENCY OF DOSING

Establishment of routes of exposure, duration, and frequency of dosing in experimental protocols requires individual attention to the objectives of the experiment. Thus, further discussion on experimental set-up involving dosing and administration is presented in subsequent chapters.

5.5.2 TOXICITY INDICATORS

The biological disposition and effect of a chemical administered to an animal test subject are determined by the toxicokinetics of the interaction between the animal and the chemical. Figure 5.1 illustrates an overview of toxicokinetic (or pharmacokinetic) disposition of a chemical *in vivo* and its purported metabolic fate.

At any time point after administration, the interaction can be identified within any compartment. In addition, depending on the influence of the chemical on the compartment, the reactions may be concurrent and/or sequential. Also, the alteration of the compartment by a toxicant influences subsequent interactions and the detection of an effect at a particular time point is shaped by the possibility of numerous events occurring between the chemical and physiologic targets. The ability to detect the effect must be responsive enough to monitor the toxicological events at that moment. Therefore, in order to apply quantitative monitors to toxicological studies, sensitive toxicity indicators within a reliable dosage range are to be instituted. Proper experimental planning requires anticipation of the toxic effects and the most suitable toxicity endpoints to monitor that effect.

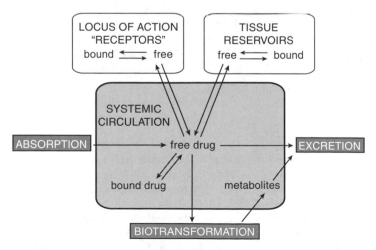

FIGURE 5.1 Overview of toxicokinetic or pharmacokinetic disposition of a chemical *in vivo* and its purported metabolic fate.

TABLE 5.3
Descriptive Animal Techniques Used as Specific Target Organs or Objectives

Method	Descriptive Protocol	Preferred Species
Dermal testing	Local irritancy	Guinea pigs
Draize method	Ocular irritancy	Rabbits
Fertility and reproductive tests	Developmental toxicity	Rabbit and rodent litters
Immunotoxicity studies	Toxicity related to immune function	Rodents
Mammalian mutagenesis	Altered foci induction in rodent liver; skin neoplasm, pulmonary neoplasm; breast cancer induction	Rodents
Mammalian teratology, whole embryo culture	Teratogenicity	Rabbits, rodents
Perinatal and postnatal studies	Developmental toxicity	Rabbits, rodents
Toxicokinetic studies	Influence on absorption, distribution, metabolism, elimination	Rodents and larger mammals

As described for *in vitro* techniques (see Part III), a battery of predefined, standardized, and validated tests is also essential for animal testing throughout the experiment. A variety of endpoints, dosages (or concentrations), and responses (effects) allow construction of a dose–response relationship that accurately reflects the events. Accordingly, a reasonable cost-effective approach to selecting appropriate toxicity indicators and inclusion of a proper number of animals will generate productive results. Data obtained from several indicators will help to construct a valid explanation of toxicological activity.

Table 5.3 outlines additional animal toxicity methods that are used along with the LD_{50} protocol or as specific target tests. These include methods for the determination of ocular irritation, local irritation, teratogenicity, developmental toxicity, perinatal and postnatal toxicity, mutagenicity, immunotoxicity, and toxicokinetics. Extensive discussion of these methods and other toxicity indicators designed for specific project objectives are discussed in subsequent chapters. Also, toxicokinetic parameters are discussed in Chapter 3.

SUGGESTED READINGS

Bayne, K.A., Environmental enrichment of nonhuman primates, dogs and rabbits used in toxicology studies, *Toxicol. Pathol.*, 31, 132, 2003.

Dirks, A.J. and Leeuwenburgh, C., Caloric restriction in humans: potential pitfalls and health concerns, *Mech. Ageing Dev.*, 127, 1, 2006.

Heilbronn, L.K. and Ravussin, E., Calorie restriction and aging: review of the literature and implications for studies in humans, *Am. J. Clin. Nutr.*, 78, 361, 2003.

Keenan, K.P., Laroque, P., and Dixit, R., Need for dietary control by caloric restriction in rodent toxicology and carcinogenicity studies, *J. Toxicol. Environ. Health. B Crit, Rev.*, 1, 135, 1998.

U.S. Department of Agriculture, National Agricultural Library, Animal Welfare Information Center, http://www.nal.usda.gov.

REVIEW ARTICLES

Barrow, P.C., Reproductive toxicology studies and immunotherapeutics, *Toxicology,* 185, 205, 2003.

Bucher, J.R., The National Toxicology Program Rodent Bioassay: designs, interpretations, and scientific contributions, *Ann. NY Acad. Sci.,* 982, 198, 2002.

Faulks, S.C, Turner, N., Else, P.L., and Hulbert, A.J., Calorie restriction in mice: effects on body composition, daily activity, metabolic rate, mitochondrial reactive oxygen species production, and membrane fatty acid composition, *J. Gerontol. A Biol. Sci. Med. Sci.,* 61, 781, 2006.

Harper, J.M., et al., Stress resistance and aging: influence of genes and nutrition. *Mech. Ageing Dev.,* 127, 687, 2006.

James, M.L., IACUC training: from new-member orientation to continuing education, *Lab. Anim. (NY),* 31, 26, 2002.

Leakey, J.E.A., Seng, J.E., and Allaben, W.T., Influence of body weight, diet, and stress on aging, survival and pathological endpoints in rodents: implications for toxicity testing and risk assessment, *Reg. Res. Perspectives,* 4, 1, 2004.

Merry, B.J., Oxidative stress and mitochondrial function with aging the effects of calorie restriction, *Aging Cell,* 3, 7, 2004.

Meyer, O., Testing and assessment strategies, including alternative and new approaches, *Toxicol. Lett.,* 140, 21, 2003.

Pauluhn, J., Overview of testing methods used in inhalation toxicity: from facts to artifacts, *Toxicol. Lett.,* 140, 183, 2003.

Pitts, M., A guide to the new ARENA/OLAW IACUC guidebook, *Lab. Anim. (NY),* 31, 40, 2002.

Rogers, J.V. and McDougal, J.N., Improved method for *in vitro* assessment of dermal toxicity for volatile organic chemicals, *Toxicol. Lett.,* 135, 125, 2002.

Silverman, J., Animal breeding and research protocols: the missing link, *Lab. Anim. (NY),* 31, 19, 2002.

Smialowicz, R.J., The rat as a model in developmental immunotoxicology, *Hum. Exp. Toxicol.,* 21, 513, 2002.

Steneck, N.H., Role of the institutional animal care and use committee in monitoring research, *Ethics Behav.,* 7, 173, 1997.

Stephenson, W., Institutional animal care and use committees and the moderate position, *Between Species,* 7, 6, 1991.

Whitney, R.A., Animal care and use committees: history and current policies in the United States, *Lab. Anim. Sci.,* 37, 18, 1987.

Section II

Toxicology Testing In Vivo

6 Acute Toxicology Testing

6.1 OBJECTIVES OF ACUTE TOXICOLOGY TESTS

The information derived from acute animal toxicology studies is essential for determining the potential toxicity of a chemical to humans and other life forms. The objective of acute studies is to identify the potential toxicology of chemicals, whether commercially available or in development. The underlying premise of using animals in acute testing is that the effects of administration of a synthetic agent on an animal mimic the possible outcomes that the agent may produce in humans or other mammals. The system also implies that short-term tests in animals parallel acute exposure in the human population.

In general, the time of exposure, dose of chemical, selection of animal species, and biological parameters used in acute studies are flexible. To qualify as an acute test, a study must be organized for comparison to human situations. The substance is administered orally, locally, parenterally, or via inhalation, and toxic effects should correlate with increasing doses. In fact, acute toxicology studies are designed to express the potency of a toxicant as a correlative dose–response relationship.

6.2 LD$_{50}$ AND ACUTE TOXICOLOGY TESTS

In accordance with U.S. Food and Drug Administration (FDA) regulations that establish current chemical and drug testing protocols using animal experiments, the LD$_{50}$ (median lethal dose 50%) of a substance must be established. The LD$_{50}$ represents the estimated dose that causes death in 50% of the population of the species exposed under the defined conditions of the test. Each LD$_{50}$ test performed for a chemical must include at least two routes of exposure, usually oral and parenteral routes. Depending on the nature of a substance, the routes can be modified to include testing for inhalation, dermal, or other selective exposure. For inhalation or aquatic studies, toxicology is presented as the median lethal concentration (LC$_{50}$) the estimated concentration of environmental exposure resulting in 50% mortality of the population of experimental animals.

LD$_{50}$ data represent lethality and do not reflect the acute toxic properties of a compound, nor does the value suggest enough information to adequately categorize a compound. The LD$_{50}$ also does not correlate well with information on different mechanisms of action of toxic agents, especially when the agents are in different toxicological categories. Comparisons are significant only when agents are homologous and have the same or similar mechanisms of action. In fact, analysis of LD$_{50}$

data simply translates a value into a comparative indicator of immediate toxicology for a specific agent in a particular strain, age group, and sex of a species of animal.

The concept of the LD_{50} developed in the early years of the 20th century when many medicinal agents were available as impure and frequently toxic mixtures, extracts, or tinctures of biologically derived materials. The test was used to standardize the potencies of such biologically derived medicinal agents. Calculation of the lethal potency of a substance could theoretically aid in assessing its therapeutic potency. Consequently, LD_{50} methodology developed to yield dose–response relationships to standardize biological preparations according to lethal potency.

In general, assessment of the lethal properties of a chemical (and hence the determination of its LD_{50}) has lost its integral value to the acute toxicology phase of the safety evaluation process. The value and significance of the LD50 determination for any agent environmental, commercial, or clinical yields little information toward explaining its mechanism of toxicity or action. The reasons for performing LD_{50} protocols are no longer valid, considering the current state of advanced biotechnology. Knowledge of the lethal range of a chemical in a particular species under specific test guidelines, does not necessarily imply protective safeguards in any other species, including humans.

It is more important to know the subtle nuances of chemical injury resulting from diverse exposure routes, injurious doses, and duration of exposure. In addition, much information about gross adverse reactions and wide ranging doses of a chemical needed to induce such reactions is ascertained from structure–activity relationships when known. Proper protective and security measures are still based almost entirely on other mechanistic toxicology tests in order to ensure safe handling, transport, and treatment upon exposure. More valuable information on the pathophysiology of toxic exposure, therefore, can be obtained from conducting other animal studies and alternative acute toxicology tests as screening models.

Consequently, it is understandable that controversy has arisen over the use of the LD_{50} as an antiquated measure of toxicology. The test originated when other indicators of toxicity were not available or were not very sensitive, and the fledgling biotechnology was not supportive of needs. In addition, the number of animals required for validation of the protocol is large and it has been suggested that the method needlessly destroys large numbers of animals. Some suggestions for updating the test include (1) the use of far fewer animals in estimating starting doses for further testing (see Section 6.4.1), (2) consideration of fewer routes of exposure, and (3) obtaining additional information during a study that adds to the body of knowledge available for the chemical. In addition, the tremendous amount of acute, toxicological data already in existence about currently available chemicals and chemical classes does not support the repeated generation of LD_{50} data with the introduction of a new formulation of the same compound, a new derivative of the same chemical class, or a different combination of known chemicals. In fact, complete elimination of the test has also been suggested.

The European Union has mandated that the LD_{50} must be phased out for all chemicals by the year 2009 (Directive OECD 2001). In addition, as animal tests are further refined and alternative tests are developed, it is anticipated that LD_{50} testing will be only of historical interest.

6.3 ORGANIZATION OF STUDIES

Historically, the selection of laboratory mice or rats for study depends more on practical considerations than on biological imperatives or applicability to humans. Rodent species are more economic to use than other animals, have fewer interspecies variations, are readily available, and are technically easier to handle. In addition, housing and caring for rodents are more feasible and practical than they are for other species.

Non-rodent LD_{50} data is necessary when the mouse and rat LD_{50} values are significantly different. Although this suggests species differences in the absorption, distribution, biotransformation, or elimination of a toxicant, this information is obtained from mechanistic studies rather than through the performance of additional LD_{50} tests. The disadvantages associated with non-rodent species are precisely the factors that discourage use of other non-rodent animal species: cost, handling, and technical measures.

Protocols for acquiring and acclimating animals are important to ensure study success. Animals should be purchased from dependable breeders and placed in the quarantine area of an animal care facility for 7 to 14 days prior to the initiation of an experiment. They are examined at appropriate stages for gross pathology and general health. A protocol includes the euthanasia of a randomly selected subpopulation of animals for blood chemistry tests and gross and histopathological examinations of organs. Extra animals are usually purchased because the screening process necessitates elimination of some animals. Acclimation to the animal care facility is important to reduce the stress associated with transport and adaptation to the new environment.

Table 6.1 summarizes other organizational considerations in toxicology studies. These include the plans associated with scheduling and performance before, during, and after the study. In general, intra- or inter-laboratory reproducibility of data depends on proper control of factors before initiation of the protocol (Table 6.1), as much as on the technical performance during an experiment.

6.4 RANGE-FINDING TESTS

Determination of LD_{50} values is not inherently precise. The original procedure first described by Trevan in 1927 often used 50 animals or more and was a benchmark for comparing the toxicologies of chemicals. The nature of the LD_{50} is that a certain percentage of animals survive a particular dose; the next higher dose also shows a percentage of animals surviving, albeit lower. The incongruity arises with the survival of animals that should not necessarily survive after they receive subsequently higher doses. Thus, the possibility exists of killing one animal with a lower dose and failing to kill another animal with the next higher dose. To avoid such anomalies, dosages must be reasonably close to demonstrate precise lethal concentrations.

Alternatively, arbitrary selection of an appropriate concentration of toxicant is administered to few animals, followed by careful observation after 24 hr. Subsequent groups of animals receive 0.75- to 1.25-fold of the original dose, depending on whether the concentration was lethal or well tolerated, respectively. Administration is further adjusted as the experiment progresses and can minimize the number of

TABLE 6.1
Necessary Steps for Procuring and Handling Animals Prior to Initiation of Experiment

Organization	Consideration	Description
Prior to purchasing	Selection of species, numbers, sexes, and ages of animals	Biological and economical calculations; applicability to human risk assessment; different or same sex; housing requirements; extra animals
	Selection of commercial animal source	Reputation of breeder, cost, availability of animals, technical support
	Type of study	Acute, chronic, target organ toxicology
	Field of study	Basic or applied toxicology testing
Immediately after purchasing and prior to initiation of experiment	Quarantine and screening	Acclimation of animals to new environment (animal care center); observe for general health or pathology; reduce stress
	Physical examination	Gross observations; blood chemistry and histopathological analysis
	Preparations	Procurement of supplies and technical help; ensure housing requirements
Immediately before initiation of experiment	Preparations	Preparation of solutions, special diets, perishable supplies; proper allocation of time
	Physical examination	Removal of unhealthy animals
During experiment	Maintain constant experimental conditions	Ensure adequate supplies, technical help; maintain housing requirements
		Maintain consistent and constant experimental protocols

animals sacrificed. A variety of established techniques have been traditionally used for LD_{50} determinations with the intention of minimizing the number of animals. These include the:

1. Up-and-down procedure (UDP) or staircase method
2. Fixed-dose approach (FDP, British Toxicology Society)
3. Acute toxic class method (ATCM)

6.4.1 UP-AND-DOWN PROCEDURE (UDP)

The revised UDP test method (ICCVAM, NIH Publication 02-4501) includes three components:

1. A primary test providing an improved estimate of acute oral toxicology with a reduction in the number of animals used when compared to classical methods

2. A limit test for substances anticipated to have minimal toxicologies
3. A supplemental test to determine slope and confidence interval (CI) for the dose-response curve

In the revised UDP primary test, an appropriate dose is administered orally to one animal (usually 175 mg/kg as the default starting dose) and the animal is observed up to 14 days. If it survives 48 hr after treatment, a preset higher dose (0.5 log spacing) is administered orally to a second animal. If the first animal dies, the second animal is dosed at a preset lower dose. Dosing stops when one of three criteria is satisfied*, with as few as six but not more than fifteen animals used per test.

In the revised UDP limit test, one animal receives a limit dose (2.0 or 5.0 g/kg). If the animal dies, the UDP primary test is conducted. If the animal lives, two more animals are dosed concurrently at the limit dose. If both of these animals live (i.e., three animals survive), the UDP limit test is stopped. If one or both of the two animals die, additional animals are dosed sequentially at the limit dose until either three animals survive or three animals die, with the maximum number of animals tested set at five. If three animals survive, the LD_{50} is above the limit dose. Conversely, if three animals die, the LD_{50} is below the limit dose level.

In the UDP supplemental test for determining slope and CI, three treatment schedules at increasing dose levels are initiated at 10- to 30-fold below the estimated LD_{50} obtained in the primary test. Dosing continues in each sequence until an animal dies. All data, including those obtained in the primary test are then considered in a statistical model that estimates the slope and CI for calculating the LD_{50}.

6.4.2 FIXED DOSE PROCEDURE (FDP)

The FDP was first proposed in 1984 by the British Toxicology Society as an alternative to the conventional LD_{50} test (OECD Test Guideline 401) for determining acute oral toxicity. The FDP used fewer animals and caused less suffering than the LD_{50} test and provided information on acute toxicology that allowed substances to be classified according to the European Union hazard classification system. In 1992, the FDP was introduced as OECD Test Guideline 420. In 1999, as part of an initiative to phase out Test Guideline 401, a review of the FDP was undertaken, the aim of which was to provide further reductions and refinements, and classification according to the criteria of the United Nations' Globally Harmonized System (GHS) of classification and labeling of chemicals.

6.4.3 ACUTE TOXIC CLASS METHOD (ATCM)

Traditional methods for assessing acute toxicology use deaths of tested animals as intended endpoints. The ATCM (OECD Test Guideline 423) avoids this criterion as

---

* Criteria include: 3 consecutive animals survive at the upper bound dose, or, 5 reversals occur in any 6 consecutive animals tested, or at least 4 animals follow the first reversal with the likelihood ratios exceeding the statistical critical value.

an exclusive indicator by incorporating evident clinical signs of toxicity at one of a series of fixed dose levels on which to base classification of the test material. Refinements to this protocol have been introduced in order to minimize suffering and distress of test animals. The oral ATCM uses three animals of one sex per step, thus incorporating considerably fewer animals. The method provides information on the hazardous properties of a chemical and allows it to be ranked and classified according to the GHS classifications of chemicals that cause acute toxicology. Another specific ATCM procedure, known as the inhalation ATC method is based on the specific requirements for inhalation toxicity testing.

The ATCM is based on a stepwise procedure, such that sufficient information is obtained on the acute toxicology of a test substance to enable its classification at any of the defined concentrations. The test requires simultaneous testing at two steps using three animals of one sex per step. If evidence demonstrates that one gender is more susceptible, the test is continued with the most susceptible gender. The absence or presence of compound-related mortality of the animals exposed at one step will determine the next step:

1. No further testing needed
2. Testing of the most susceptible gender only
3. Testing of an additional three male and three female animals at the next higher or next lower concentration level

When there are indications that a test material is likely to be non-toxic, a limit test is performed as described above. In addition, the ATCM further outlines standards and recommendations for the selection of animal species, housing and feeding conditions, preparation of animals, and mode of exposure, among other proposals.

6.5 CLASSICAL LD$_{50}$

A variety of factors should be considered when planning and organizing a study for the determination of LD$_{50}$, not all of which are associated with the calculation of 95% confidence limits. Most preparations among the different routes of administration are similar although some changes are required as shown below.

6.5.1 Oral LD$_{50}$

Table 6.2 summarizes some of the parameters and factors that are monitored as a classical oral LD$_{50}$ study commences. Such factors include but are not limited to (1) randomization of animals, (2) maintenance of a narrow range of body weights, (3) appropriate number of animals per group, (4) identification of individual test subjects, (5) fasting, and (6) availability of water. Dosage range is determined according to techniques outlined above. Preliminary range-finding experiments are initiated to minimize the extent of no lethality or 100% lethality, thus reducing the number of groups in the total study. This method also improves the precision of the LD$_{50}$ determination.

TABLE 6.2
Considerations during Preparation for Oral LD$_{50}$ Study

Parameter	Factors for consideration
Randomization of animals	Unbiased distribution into groups
Narrow range of body weights	Uniform distribution of similar sized animals
Number of animals	For classical LD$_{50}$, average of 10 per treatment group
Identification of individual animals	Ensures individual observation and monitoring; allows for group housing
Fasting (16 to 24 hr)	Optimal GI absorption
Water *ad libitum*	Prevention of dehydration

TABLE 6.3
Considerations during Preparation for Dermal LD$_{50}$ Study

Parameter	Factors for consideration
Formulation	Solids dissolved in water or inert oil-based vehicle (dimethyl sulfoxide, propylene glycol); liquid, paste, ointment, patch
Application to skin	Shave fur 24 hr prior to test; uniform application; 2 to 3 cm^2 area for smaller, 3 to 5 cm^2 for larger animals (10%); consistency; volume = 1 to 2 ml/kg
Absorption of toxicant	Depends on water-soluble or lipid-soluble properties (more rapid)
Solubility	Lipid-soluble liquid more rapidly absorbed; water-soluble liquid requires more time; occlusive dressing used if necessary
Variability of results	Dermal studies associated with high degree of variability; determine LC$_{50}$

6.5.2 DERMAL LD$_{50}$

Dermal LD$_{50}$ studies are conducted on toxicants if the probable exposure route is through skin absorption. As with the oral LD$_{50}$, lethality is generally assessed in two species, one of which is non-rodent. Also, the test substance is applied to shaved skin in increasing doses to several groups of experimental animals, one dose per group. The parameters involved with the dermal LD$_{50}$ determination are the same as those described for the oral LD$_{50}$ (Table 6.2). Some of the factors that are unique to dermal studies are summarized in Table 6.3.

Most of the variability in dermal LD$_{50}$ studies arises from these parameters. Incomplete absorption of a toxicant due to poor vehicle solubility, inability to penetrate intact skin, and lack of uniformity in the application method are largely responsible for the inconsistencies seen with dermal LD$_{50}$ studies. Because of known or suspected inadequate absorption, an LC$_{50}$ (median lethal concentration 50%) is determined after absorption and may substitute for the lack of a reliable dosage determination. In fact, cut-off values of 1 g/kg doses constitute an upper limit. Also, animals are prepared at least 24 hr prior to initiation of the experiment, and the application area is prepared to encompass about 10% of total body surface area. Application procedure and volume of material is consistent and uniform — planned

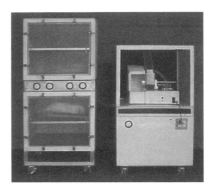

FIGURE 6.1 Photo of rodent exposure chamber for inhalation (smoking) toxicology studies. Shown is the TE-10 smoking machine, hood unit, air compressor, rat exposure chambers, meters, and controls. (Courtesy of Teague Enterprises, Davis, California.)

so as to minimize technical variability. Occlusive devices or mechanical restraints are avoided if possible.

6.5.3 INHALATION LD$_{50}$

Air-borne toxic materials that are transported via gases, aerosols, smoke, or ventilation necessitate the determination of acute inhalation LD$_{50}$. As described for the classical LD$_{50}$, rodents and a non-rodent species are exposed for 4 to 24 hr to a test substance in increasing concentrations (one concentration per group, at least four doses plus a control group). Well controlled inhalation studies incorporate a negative pressure, dynamic inhalation system with programmable airflow settings. Currently used systems, such as that shown in Figure 6.1, are capable of delivering precise test material concentrations, continuously monitoring toxicant in the exposure chambers, and include exhaust ducts capable of shunting chamber air through ductal fume vents. Some systems accommodate either whole body exposure or oral–nasal exposure only. The range of doses is capable of producing a corresponding series of toxic effects and mortality rates to facilitate assessment of acute toxicity for LD$_{50}$ or LC$_{50}$ determination.

Table 6.4 presents some of the difficulties unique to inhalation studies. The concentration of toxicant is calculated as the total amount of test substance delivered through the inhalation system with respect to the air pressure and volume of air, as monitored by head-space gas chromatography. Concentrations are calculated, monitored, and adjusted as necessary. However, net toxic concentrations delivered to the animals are influenced by airflow rates, particle size distribution, and air temperature and humidity, all of which can be monitored via serial programmable detectors.

6.6 OTHER CONSIDERATIONS WITH LD$_{50}$ DETERMINATIONS

Adequate planning of an experimental protocol minimizes errors and unanticipated results. Consideration of several factors, as discussed above, reduces experimental

TABLE 6.4
Parameters for Consideration — Special Emphasis on Inhalation LD$_{50}$ Studies

Parameter	Factors for Consideration
Concentration of delivered test agent	Airflow rate into chamber; air temperature and humidity; monitoring concentration in chamber; check integrity of exposure chamber
Particle size	Particle size of particulates determines distribution to target organ (lower or upper respiratory tract); size influences solubility of toxicant in air
Respiration rate	Light and dark cycles affect respiration rate; contact with irritant induces behavioral, secretory, and inflammatory responses altering exposure rate
Control groups	Influence of co-solvents, additives, air pressure

error, optimizes results analysis, and maximizes resource utilization. In addition, laboratories are now encouraged to consider using currently available *in vitro* toxicology screening protocols and assays in addition to *in vivo* studies. *In vitro* studies are particularly beneficial as preliminary screening tests prior to animal toxicity studies. They are also used to supplement, reduce, or refine animal testing as well as yield mechanistic data in support of LD$_{50}$ determinations.

6.6.1 ROUTE OF ADMINISTRATION

A toxic agent is administered according to the environmental, occupational, or clinical exposure expected in humans or animals. Although most exposures are via the oral or local route (and not limited to ocular, dermal, or pulmonary routes) as encountered with occupational exposure, some agents such as therapeutic drugs and biologicals may require testing by parenteral routes. Oral administration may also be reflected in animal testing using several methods. For instance, acute oral administration is accomplished through:

1. Adding a chemical to drinking water. This requires controlled distribution of the volume of water to deliver the appropriate concentration of test chemical.
2. Adding a test chemical to solid food. This also necessitates monitoring of food volume intake and includes the risk of reduced absorption of chemical in the presence of stomach contents.
3. Administration via gastric gavage. The agent is dissolved in a solvent vehicle and administered through a round tipped gastric needle and syringe. By maintaining constant volumes (2 to 5 ml/kg total volume per oral injection), precise dosages (mg/kg body weight) are allotted.
4. Diet supplement. Animals have natural aversions to otherwise miscible liquid chemicals such as alcohols and bitter, irritating, organic liquids that are added directly to drinking water, to the extent that the animals will dehydrate before consuming enough of the test substance. In this situation, a completely liquid diet is prepared from commercially available fortified liquid diets or reconstituted powder formulations. The liquid diet is then

supplemented with the active ingredient by volume (% w/v), allowing for well controlled delivery of test material. After initial acclimation and weight loss, the combination of hunger and thirst drives overcomes the natural aversion to the added test ingredient.

Dermal applications of toxic substances are also subject to solubility considerations such as dissolution of the active ingredient in a solution, powder form, ointment, cream, or paste. The objective is to ensure adequate penetration of the vehicle and active ingredient through dermal barriers. This may necessitate covering the treated area with occlusive wrapping. The substance remains in contact for the specified exposure time, after which the area is rinsed clean of material.

With inhalation exposure of chemicals in the form of gases, vapors, particulate matter, or aerosols, the principal consideration is the delivery of the correct concentration. In addition, inhalation studies require placing adequate numbers of animal in a toxicant-infused chamber at different times to accommodate the range of concentrations. Exposure times vary from 2 hours up to 96 hr, after which the animals are removed from the chamber.

6.6.2 DURATION

Although LD_{50} studies are generally performed during the first 24-hr exposure period, needless suffering is avoided by removing and euthanizing prostrate, ill, or dead animals from treatment groups. Sufficient results are also obtained by determining the time of morbidity or mortality, as well as employing any toxicological analyses.

6.6.3 GENERAL APPEARANCE OF ANIMALS

Although mortality is the major goal of an LD_{50} study, data gathering may be optimized by performing other toxicological analyses. Close observation of treated animals throughout a study period should include:

1. General appearance (compared to control animals)
2. Onset, intensity, and duration of toxic effects
3. Changes in behavior, activity, respiration, appetite, fluid intake, or food retention
4. Monitoring of body weight, food intake, water consumption, and skin or fur turgor

6.6.4 SPECIMEN COLLECTION AND GROSS PATHOLOGY

Depending on the animal species, blood samples are obtained during the course of an experiment and at its conclusion, from both rodent and non-rodent species. Microliter blood samples are acquired using puncture methods or small skin incisions. Milliliter blood samples are secured using tail or ocular venipuncture in rodents or ear venipuncture in larger mammals. In addition, urinalysis and fecal analysis are readily accomplished on samples collected from animals placed overnight in appropriate metabolic cages. Non-invasive neurological assessment is performed with behavioral testing apparatus (rotorod, maze, and learning reinforcement activities). In addition, the study is designed to conduct gross pathological examination at any time point, including histological examination of tissue specimens and hematological analysis. Selected organs are quickly and precisely dissected into 1- to 2-cm sections and immersed in an appropriate fixative for further histological processing.

6.6.5 BIOLOGICAL VARIATION

Intra- and inter-species differences exist in the metabolism and toxicology of chemicals and within a population of animals randomly selected for a study. Variation of response is observed in heterogeneous, randomly bred, strains of animals, posing the possibility of encountering substantial errors in estimating LD_{50} values. In addition, significant skewing of the dose–response relationship may be encountered in inbred animals — animals whose homogeneous genetic pool origins suggest similar responses.

In addition, extrapolations of LD_{50} data to human toxicology and carcinogenicity are not necessarily linear because humans may be more susceptible than monkeys, rats, and mice to toxicological effects in general and carcinogenic chemicals in particular. Striking inter-species variation in susceptibility to several carcinogens is known; for example, rats are the most sensitive species to aflatoxin B1. Efficient conjugation with glutathione, however, confers aflatoxin B1 resistance to mice. Extremely large inter-species differences in herbicide and pesticide-induced toxicity are known among guinea pigs (more susceptible) and hamsters (most resistant). One of the most complex challenges, therefore, is awareness of the biological variations present within species and extrapolation of this information from laboratory animals to humans.

6.6.6 DETERMINATION OF ACUTE LETHALITY

As noted in Chapter 3, the path by which a dose elicits a particular response is known as the dose–response (or concentration–effect) relationship. The observed response is a calculated observation, assuming that the response is a result of exposure to a chemical and is measured and quantified. The response also depends on the quantity of chemical exposure and administration within a given period. Two types of dose–response relationships exist, depending on number of subjects and doses tested. The *graded dose–response* (Figure 6.2) describes the relationship of test subjects to logarithmic increases in the dose or concentration of a chemical. The concentration of the chemical is proportional to the number of surviving subjects in the experimental system or any other parameter of morbidity.

The *quantal dose–response* (Figure 6.3) is determined by the distribution of responses to increasing doses in a population of test subjects. This relationship is generally classified as an all-or-none effect and the animals are quantified as either responders or non-responders. Figure 6.4 shows a typical graph for comparing ED_{50} (median effective dose 50%) to LD_{50}. Because the LD_{50} is a statistically calculated dose of a chemical that causes death in 50% of the animals tested, it is an example

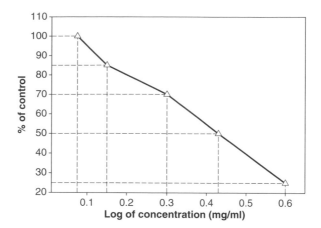

FIGURE 6.2 The graded dose–response (or concentration effect) curve for a chemical administered over a set time period. For LD_{50} studies, the measured parameter (percent of control) is mortality. The diagram illustrates extrapolation of the LD_{50} (0.43 mg/ml), as well as the LD_{75} (0.6 mg/ml), LD_{30} (0.3mg/ml), and LD_{15} (0.15 mg/ml). The latter represent cut-off values corresponding to 25, 70, and 85% of controls. (From Barile, Frank A., *Clinical Toxicology: Principles and Mechanisms,* CRC Press, 2004.)

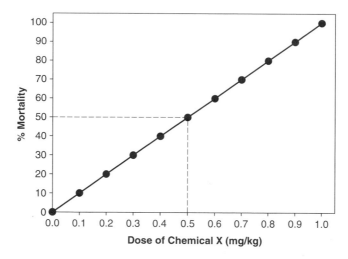

FIGURE 6.3 Quantal dose–response curve showing experimental derivation and graphic estimation of LD_{50}. (From Barile, Frank A., *Clinical Toxicology: Principles and Mechanisms,* CRC Press, 2004.)

of a typical quantal dose–response curve. The doses administered are continuous or at different levels and the response is generally mortality (although gross injury, tumor formation, or other measurable criterion is used to determine a standard deviation or cut-off value).

Graded and quantal curves are generated based on several assumptions. The time at which the response is measured is chosen empirically or selected according to

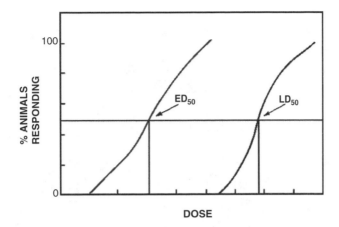

FIGURE 6.4 Representative dose–response curve for median effective dose (ED_{50}) and median lethal dose (LD_{50}). (From Barile, Frank A., *Clinical Toxicology: Principles and Mechanisms,* CRC Press, 2004.)

accepted toxicological practices. For instance, empirical time determinations may be established using a suspected toxic or lethal dose of a substance and the response is determined over 24 to 96 hr. This time period is then set for all determinations of LD_{50}. The frequency of administration is assumed to be a single dose administered at the start of the period when the test subjects are acclimated to the environment.

Another assumption in the determination of the LD_{50} is that the observed effect is, in fact, due to the presence of the chemical. The establishment of this causal relationship is critical if valid conclusions are drawn from the dose–response curve. It is also presumed that the chemical in question is present at the receptor site or molecular target affected by the substance. Support for this assumption follows from measurement of concentrations of the test chemical at organ level or in the plasma. In fact, as the concentration of a chemical in the affected compartment increases, the degree of response must increase proportionately if this assumption is valid (hence the derivation of the *concentration effect* term).

Determination of the LD_{50} further entails that the response to a chemical is normally distributed — the highest number of respondents are gathered in the middle dosage range. Figure 6.5 represents a normal frequency distribution achieved with increasing doses of a chemical versus the cumulative percent mortality. The bars represent the percentage of animals that died at each dose minus the percentage that died at the immediately lower dose. As shown by the normal (Gaussian) distribution, the lowest percentage of animals died at the lowest and highest doses, accounted for by biological variation.

Consequently, the calculation of the LD_{50} considers the slope of the line with respect to response and the 95% confidence limits. The slope is an indication of the range of intensity of toxicity of the chemical in question. For instance, a steep slope represents proportionately greater responses to increases in concentrations, whereas a shallow slope suggests less toxic responses in proportion to increases in dose. The confidence interval is a relative measure of the degree of error present in the

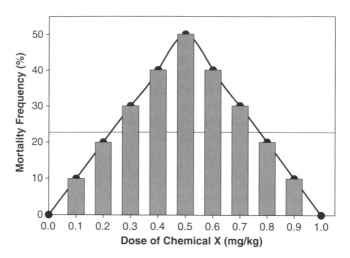

FIGURE 6.5 Normal frequency distribution of mortality frequency (%) versus dose. (From Barile, Frank A., *Clinical Toxicology: Principles and Mechanisms,* CRC Press, 2004.)

population sample and allows for intra- and inter-laboratory comparisons. Statistically, LD_{50} determinations, percent response computations, determination of confidence intervals, regression analysis, and correlation coefficient, all of which are integral components of an LD_{50} calculation, are further described in Chapter 8.

6.7 APPLICATIONS OF LD_{50} STUDIES

Because the LD_{50} value is a statistical estimate of the acute lethality of a chemical administered under specific circumstances, it provides a measure of relative toxicities of chemicals under similar or identical conditions. Thus the major application of the LD_{50} is comparative, allowing for semi-quantitative toxic evaluations of compounds, especially within the vast database of acute toxicology recorded for laboratory rodents. In addition, the test provides a screening method for toxic evaluation, particularly useful for new unclassified substances. The determination, however, is not without limitations. By current scientific methodologies, the LD_{50} is antiquated, requires large numbers of animals, does not provide significant information regarding mechanistic effects or selective target organs, and does not suggest complementary or discriminating pathways of toxicology. It is also limited by the route and duration of exposure. Consequently, its routine use in toxicology testing has become the subject of continuous debate and regulatory review.

SUGGESTED READINGS

Clode, S.A., Assessment of *in vivo* assays for endocrine disruption, *Best Pract. Res. Clin. Endocrinol. Metab.,* 20, 35, 2006.
Huxley, A., Testing is necessary on animals as well as *in vitro*, *Nature*, 439, 138, 2006.

Lamb, J.C. and Brown, S.M., Chemical testing strategies for predicting health hazards to children, *Reprod. Toxicol.*, 14, 83, 2000.

Travis, K.Z., Pate, I., and Welsh, Z.K., The role of the benchmark dose in a regulatory context, *Regul. Toxicol. Pharmacol.*, 43, 280, 2005.

REVIEW ARTICLES

Botham, P.A., Acute systemic toxicity: prospects for tiered testing strategies, *Toxicol. in Vitro*, 18, 227, 2004.

Brent, R.L., Utilization of animal studies to determine the effects and human risks of environmental toxicants (drugs, chemicals, and physical agents), *Pediatrics*, 113, 984, 2004.

Gelbke, H.P., Kayser, M., and Poole, A., OECD test strategies and methods for endocrine disruptors, *Toxicology*, 205, 17, 2004.

Krishna, G., Urda, G., and Theiss, J., Principles and practices of integrating genotoxicity evaluation into routine toxicology studies: a pharmaceutical industry perspective, *Environ. Mol. Mutagen*, 32, 115, 1998.

Meyer, O., Testing and assessment strategies, including alternative and new approaches, *Toxicol. Lett.*, 140, 21, 2003.

Pauluhn, J., Overview of inhalation exposure techniques: strengths and weaknesses, *Exp. Toxicol. Pathol.*, 57, 111, 2005.

Rispin, A., et al., Alternative methods for the median lethal dose (LD_{50}) test: the up-and-down procedure for acute oral toxicity, *ILAR J.*, 43, 233, 2002.

Rosenkranz, H.S. and Cunningham, A.R., Lack of predictivity of the rat lethality (LD_{50}) test for ecological and human health effects, *Altern. Lab. Anim.*, 33, 9, 2005.

Rosolen, S.G., Rigaudiere, F., Le Gargasson, J.F., and Brigell, M.G., Recommendations for a toxicological screening ERG procedure in laboratory animals, *Doc. Ophthalmol.*, 110, 57, 2005.

Schlede, E. et al., Oral acute toxic class method: a successful alternative to the oral LD_{50} test, *Regul. Toxicol. Pharmacol.*, 42, 15, 2005.

Stallard, N. and Whitehead, A., A statistical evaluation of the fixed dose procedure, *Altern. Lab. Anim.*, 32, 2004.

The Revised Up-and-Down Procedure: A Test Method for Determining the Acute Oral Toxicology of Chemicals, 2001, NIH Publication 02-4501, http://iccvam.niehs.nih.gov/docs/docs.htm#udp.

Walker, D.K., The use of pharmacokinetic and pharmacodynamic data in the assessment of drug safety in early drug development, *Br. J. Clin. Pharmacol.*, 58, 601, 2004.

7 Subchronic and Chronic Toxicology Testing

7.1 INTRODUCTION

As discussed in Chapter 6 covering acute toxicity studies, the shorter duration of exposure in acute studies allows monitoring of the effects of a chemical at high doses in an abbreviated time period. Such acute exposures occur in a variety of circumstances, particularly in emergency, occupational, environmental, and domestic settings. More frequently, clinical toxicology has shown that acute exposures to drugs of abuse and therapeutic drugs are of major international concern and treatment of accidental or intentional overdoses poses a significant challenge. For purposes of comparison, however, as much toxicity occurs with extended interactions with lower non-fatal concentrations of chemicals as occurs with short intense exposures.

In addition, much regulatory emphasis is placed on long-term human and animal exposures to environmental toxins, occupational exposure to hazardous substances, and chronic effects of food and dietary supplements. Consequently, there exists an important incentive for the scientific toxicology sector to develop test methods for predicting relative chronic hazards and risks of exposure to chemicals in daily life. As a result, simulation of these interactions requires the development of several different types of studies in order to mimic the human situation.

7.2 TYPES OF SUBCHRONIC AND CHRONIC TOXICITY TESTS

7.2.1 OBJECTIVES AND DEFINITIONS

In general, in animal and human risk assessment, acute and chronic durations of exposure are relative terms intended for comparative purposes. Any exposure up to 24 hr is regarded as *acute*. Exposure to many toxic gases (e.g., carbon monoxide, hydrogen cyanide) requires less than 24 hr to produce toxicity. Also, 72 hr may still constitute an acute exposure, such as in continuous low-dose exposure of children to acetaminophen and exposure of house pets to herbicides and insecticides. *Subacute* exposure is a term that has fallen into disuse. It formerly referred to repeated exposure of more than 72 hr but less than a month.

Chronic exposure is any relative time period for which continuous or repeated exposure beyond the acute phase is required for the same chemical to induce a toxic response. For convenience, the *chronic* term also applies the subchronic time period.

Subchronic exposure is also understood to involve a duration between acute and chronic. The traditional time of subchronic exposure is understood to be a period of 1 to 3 mo. Thus, all these terms are flexible adaptations for defining the onset of chemical intoxication and considerable overlap in judgment may occur in assigning labels to exposure periods. In contrast to acute studies, the objectives of subchronic and chronic studies are to:

1. Determine toxicological effects of repeated administration of test chemicals on potential target organs at subchronic or chronic dosages
2. Establish dose–response (or concentration effect) relationships using various indicators over the selected dosage ranges and durations of exposure
3. Experimentally verify a maximum dosage level that does not promote ostensible toxicity with repeated exposure
4. Propose a mechanism of toxicity that complements or contrasts with acute studies of the chemical or class of agents

The following discussion describes the information necessary for addressing these objectives, and explains how the experimental approach differs from the design of acute toxicology testing studies. It is important to note that comparability of results depends upon the similarities of the test criteria.

7.2.2 FACTORS ASSOCIATED WITH CHRONIC TOXICITY

7.2.2.1 Frequency of Exposure

It is important to emphasize that frequency of exposure involves repeated doses of a toxin during a certain time period. Thus, continuous repeated exposure to a toxin, especially during a subchronic or chronic time period, has greater toxic potential. In addition, dose, duration, frequency, and route of exposure contribute to chemical toxicity in part through accumulation of a compound in physiological compartments. A chemical's toxic profile is determined according to its half-life $(t_{1/2})$ in plasma and its effect, that is, the time required for plasma levels to decrease to one-half of the measured or estimated concentrations. If the frequency of exposure exceeds the $t_{1/2}$ of a chemical, its concentration in a compartment is likely to increase beyond the desirable level. Consequently, in long-term interactions with chemicals, accumulation results from overloading of an agent within this compartment.

7.2.2.2 Accumulation and Distribution

Once absorbed, a chemical distributes and/or binds to one or more of many physiological sites. The distribution of a chemical depends largely on its physiochemical characteristics. The compartments include whole blood, serum and serum proteins, plasma and plasma proteins, adipose tissue, interstitial and extracellular fluids, alveolar air space, and bone marrow. In addition, any tissue or organ may preferentially accumulate a chemical, thus acting as a discrete compartment, as with the preferential accumulation of heavy metals in adipose tissue.

Consequently, toxicity may be experienced for a prolonged period as a compound is slowly released from this compartment, years after exposure has ceased. Accumulation, therefore, is predicted based on a chemical's apparent volume of distribution (V_d), which is estimated as the total dose of drug in the body divided by the concentration of drug in the plasma for a given period. In general, the greater the V_d, the greater the potential for accumulation in some physiological compartment. A standard measure for accumulated internal quantity of a chemical is the *body burden*, defined as the amount of chemical stored in one or several physiological compartments or in the body as a whole.

7.2.2.3 Influence of Chemical Structure

Accumulation and its influence on chronic toxicity is also determined by a chemical's structure and its interactions within physiological compartments. This phenomenon is guided by the chemical's predominant ionic or non-ionic state of existence in a physiological fluid. In general, at physiologic pH, lipid-soluble compounds preferentially remain in a non-ionic state, preferring to bind to, penetrate, and accumulate in membranes of tissues and organs. Conversely, water-soluble compounds remain as ionic species at the pH of the blood. Thus, because they are less prone to tissue binding, the ions are readily available for renal secretion and elimination.

7.2.3 OBJECTIVES OF CHRONIC TOXICITY STUDIES

The goals of conducting chronic toxicity studies are similar to those of acute studies, with a few important differences. The objectives that overlap with acute studies include (1) determination of the lethal and toxic concentrations of a chemical and its effect on organs and tissues, (2) identification of the causal relationship between the administered dose and the altered physiological, biochemical, and morphological changes, and (3) monitoring of animal species variations in response to an agent. Major differences between chronic studies and acute experiments, however, rely on frequency, accumulation, and length of exposure to a toxic agent. For instance, chronic studies are generally conducted to:

1. Measure or assess the toxic effects of lower, more frequently administered doses of a chemical, thus analyzing for repeated cumulative exposure
2. Determine cumulative effects of repeated exposure
3. Examine the toxicological effects of increasing doses of chemicals over extended periods
4. Identify recovery of subjects after removal of the source of exposure
5. Predict long-term adverse health effects in the species arising from intermittent, repeated, or continuous exposure

Finally, chronic studies generally complement acute studies, assuming that the conditions are structured to enhance the results obtained in acute experiments.

7.3 EXPERIMENTAL DESIGN

As noted above, *chronic* and *subchronic* are often relative terms, especially in relation to the species. Historically, 2 yr is a typical period for conducting chronic studies in rodents. However, 2 yr does not represent a significant portion of the total life spans of other species such as dogs (average life expectancy of 9 to 10 yr) or rabbits (5 to 6 yr). Even in rodents sustaining calorie-restricted diets, the 2-year exposure period is significantly less representative of their healthier, longer life spans.

In general, as the life expectancy of the mammalian species increases, the exposure period must be adjusted to mimic a chronic exposure. Thus it becomes more difficult to assign and extrapolate the chronic time period in a particular species to adequately correlate the results of such a study to human risk assessment. Consequently, the lengths of chronic studies (6 mo to 3 yr) are variable and flexible and are determined according to the appropriateness of the objectives of a project.

7.3.1 Selection of Dosage Levels

Dosage levels for chronic toxicology studies are selected based on existing information available about a chemical from acute studies, known toxicological effects, animal and human epidemiological data, knowledge of the species reaction to chemicals of similar classes, and known toxic concentrations of chemicals from similar classes. In addition, *in vitro* data is also recommended as a method for screening chemicals for *in vivo* studies. Thus, familiarity and awareness of dose-response curves, LD_{50}, EC_{50}, and IC_{50} (*in vitro* inhibitory concentration 50% or concentration that causes 50% of measured response from a control in an *in vitro* system) values for the chemical aid an investigator in identifying proper doses for a chronic *in vivo* study.

Understanding the details of the target endpoint is also essential for determining dosage levels. As will be seen in later chapters, the dosages established in studies whose objectives involve the evaluation of histological or pathological monitoring for toxicity may be different from carcinogenic projects. As with acute studies, an investigator has some flexibility in the selection of dosage levels, usually numbering up to five groups plus a vehicle and/or blank control. In addition, preliminary experiments with small numbers of animals with repeated dosages are necessary to establish lowest and highest effector groups.

Doses are then selected based on preliminary range-finding pilot projects. Finally, recognition of the pharmacokinetics of the chemical, if available, contributes to the prediction of its behavior in physiological compartments during the course of a study. Consequently, a pilot study on limited numbers of animals for several weeks supplies the necessary information to commence a full study. It also reduces the chances of altering the parameters during the course of the project such as adjustment of doses, significant changes in the number of animals, or modification of the study parameters to accommodate for excessive mortality or lack of effects.

7.3.2 Selection of Animals

Selection of the number and species of animal relies on the objectives of a chronic study, the parameters established in the experimental set-up, and the ability to

TABLE 7.1
Steps Necessary for Procuring and Handling Animals Prior to Initiation of Chronic Experiments

Organization	Consideration	Description
Prior to purchasing	Selection of species, numbers, sex and age of animals	Biological and economical calculations; applicability to human risk assessment; usually same sex; long-term arrangement for housing requirements; young animals
	Selection of commercial animal source	Same as with acute studies: reputation of breeder, cost, availability of animals, technical support
	Type of study	Toxicological assessment; for carcinogenicity study
Immediately after purchasing and prior to initiation of experiment	Quarantine and screening	Same as with acute studies: acclimation of animals to new environment (animal care center); observe for general health or pathology; reduce stress
	Physical examination	Gross observations; histopathological and blood chemistry analysis
	Prepare for experiment	Procure supplies, technical help, ensure housing requirement for duration of study
Immediately before initiation of experiment	Prepare for experiment	Continuous preparation of solutions, special diets, perishable supplies
	Physical examination	Removal of unhealthy animals
During experiment	Maintain constant experimental conditions	*Essential for chronic studies*: ensure adequate supplies, technical help, maintain housing requirements
		Maintain consistent and constant experimental protocols

monitor and adjust to unanticipated and untoward results. In general, the selection of animals is as important in preparing for chronic studies as it is in acute experiments. Table 7.1 outlines some of these parameters that bear similarities to those established for acute studies (see Table 6.1, Chapter 6). However, considering that chronic experiments are longer in duration, attention is drawn to understanding the toxicological indicators and effects that may be unpredicted in such experiments.

In addition, maintenance of constant experimental conditions during the course of a 2-year study is more daunting than maintaining conditions for 24- to 96-hr acute experiments. Other factors to be monitored during the conduct of a chronic toxicity testing project are outlined in Table 7.2. Some of the circumstances that require premature sacrificing of animals are included. It is important that the decision to terminate the animal's inclusion in the experiment is documented and that enough analyses of specimens are performed to determine a causal relationship, that is, whether the state of the animal is induced by experimental conditions or by factors unrelated to the test chemical.

TABLE 7.2
Circumstances Requiring Consideration or Modification during Chronic Toxicity Study

Circumstance	Factors for Consideration
Euthanasia	Impaired health, anorexia, debility, pain, weight loss
Number of treatment groups	According to number of doses (low, intermediate, high) plus control
Number of animals per groups	5 to 10; criteria based on efficient use of data and avoiding unnecessary wasting of animals
Length of study	Range of 90-day to 3-yr length
Representative time periods for analysis during study	Subgroups of animals analyzed at selected time points during course of project; at least four intervals
Recovery study	Performed at termination of chronic study with surviving animals

Other considerations during the course of a chronic study include (1) the importance of acclimating the animals to the housing and laboratory environment prior to initiation of the study as with acute studies, (2) maintaining the animals free of infection, and (3) purchasing extra animals for smaller parallel experiments that provide a set of baseline values for physiological, biochemical, and morphological endpoints during the treatment and post-treatment periods and during parallel acute range-finding studies. Finally, as with acute studies, the animals are randomly assigned to treatment groups and cages, labeled for identification, repositioned weekly in the quarters, and examined using physiological and toxicological indicators.

7.3.3 MONITORING DAILY CRITERIA OF CHRONIC TOXICOLOGY STUDIES

Chronic studies require persistence, diligence, and endurance on the part of the investigator and personnel, principally because of the need for daily observation and handling of animals. Table 7.3 lists criteria for monitoring the daily routine aspects of a chronic study. Attention to these factors aids in analyzing daily progress of a study, ensures humane treatment of animals, and strengthens the reliability of results.

TABLE 7.3
Monitoring Routine Features of Chronic Studies

Routine Feature	Observations
Appearance	Condition of fur, general cleanliness
Behavior	Grooming, alertness, aggressiveness, irritability, irascibility (ease of excitement), food and water consumption, sedation
Cages	Removal of waste, absence of foul odors, fresh bedding
Quarters	Adequate lighting, proper day-and-night cycles, temperature and humidity control, weekend and holiday schedules
Animal care personnel	Adherence to schedule, sufficient coverage for off-hours, training

TABLE 7.4
Monitoring Biological Parameters during Chronic Studies

Indicator	Biological Parameters Measured
Observation for moribund or dead animals	Necropsy, organ/body weight ratio, fixation and processing of specimens for histological analysis
Time	Documenting appearance of pathology, sampling intervals, specimen collection, time of death
Aging	Differentiating age-related changes from agent induced toxicity
Euthanasia	Interval sampling of selected groups of animals for intra-study monitoring
Normal physiological functions	Food and water intake, pair feeding,* urine output, fecal excretion
Physiological parameters	Body weights, measurement of daily food and water consumption, blood and urine clinical chemistries

* Pair-feeding refers to matching groups of treated and untreated animals; both groups receive the same quantities of nutrition. Assuming that the chemical treatment alters adequate food and liquid intake, the treated group thus becomes the rate-limiting nutrition control while the control group serves as an indicator for differentiating the effect of the toxicant from that of inadequate nutrition.

In addition, daily observations aid staff in noticing subtle underlying toxic phenomena or target organ pathology.

Daily observation of animals also involves minding routine biological parameters as indicators of health status. Table 7.4 outlines biological indicators that are employed during selected intra-study intervals to ascertain biological condition. As with the routine observations, physiological and morphological parameters are designed to assess appearance and time sequence of toxicity, especially in the detection of target organ toxicity. In addition, general criteria used to determine daily appearances of the animals may overlap with monitoring the same criteria, for example, observing when and how an animal approaches the food and water versus measuring the amount of nutrition consumed (Table 7.3 and Table 7.4). This is particularly important if the chemical additive interferes with appeal for the solid or liquid nutrition. Consequently, careful observation of dietary habits avoids the confusion between toxicity due to chemical agent versus the incorrect conclusion of the development of a pathology due to inadequate dietary intake.

Blood, urine, and fecal specimen collection is performed at selected intervals without the need to destroy animals prior to termination of the study. Urine and fecal collection devices are non-invasive and are available as attachments to cages. Blood sample collection requires training but is performed either by venipuncture via the tail vein method or through capillary puncture of the inner canthus of the eye. In both situations, improper technique or repeated use of the same animal presents risk of infection. Non-invasive procedures are also used to assess neurological, behavioral, and learning parameters. These include techniques such as measuring motor rod balance, maze performance, tail-flick procedure for pain measurement, and reward stimuli for learned behaviors.

Urinalysis, blood clinical chemistry, and hematology measurements are routine indicators of toxicity and ensure identification of toxicological events. These indicators are also suggestive for mechanisms of toxicity and lead to the understanding of target organ effects. In particular, plasma levels of specific enzymes are often reflective of target organ damage and are used to identify interactions occurring within other organs. Similarly, routine urinalysis provides evidence of toxicant-induced changes and is suggestive of metabolic as well as target organ toxicity. Considerable information including evidence of biotransformation products of administered chemicals can be obtained from these measurements.

7.3.4 DURATION OF STUDIES

During the course of chronic and subchronic studies, it is important to understand that selection of a termination time period should prevent interpretation of age-related changes and pathologies as changes due to chemical induction. Subchronic studies are generally 21 to 90 days in duration, depending upon the route of administration and toxicological endpoints of interest. Time periods for the duration of chronic studies, however, have not been clearly defined.

Traditionally, based on the approximate life spans of rodents, chronic experiments have been designated to continue for 2 yr. The FDA recently suggested that these studies should be reduced to 6 to 18 mo, depending on the animal species. The basis for this argument relies on recent evidence suggesting that toxicity observed in the second year of a chronic study is similar to that seen during the first year. In fact, since age-related mortality and morbidity are uncommon in the early months of chronic experiments, most toxicities observed during this time are causal and dose-dependent.

This approach however may not adequately mimic the level of lifetime exposure for a rodent. In addition, the susceptibility of an animal to chemical insult increases with age, thus obviating pathological consequences of chemical administration in older animal populations. Consequently, it may be necessary to design separate projects whose objectives focus on detecting age-dependent, toxicant-induced pathology. These studies may need to highlight different starting ages for the animals, continuing with exposure for 6 to 12 mo. A primary concern, therefore, is that chronic studies are designed according to the expected lifespans of the species involved.

7.3.5 RECOVERY EXPERIMENTS

A recovery experiment or the reversibility of toxicant-induced injury after prolonged exposure is important for understanding cumulative toxicity, reorganization of physiological parameters, regeneration of normal biological status in an organism, and repair mechanisms. The proper design of subchronic or chronic studies should include efficient use of resources to incorporate recovery from toxicity. The inclusion of such studies in a project helps identify the progression, regression, or amelioration of toxicant-induced injury after termination of exposure. A post-treatment recovery period of 1 to 3 mo is usually adequate to address these objectives and is accomplished by random selection of representative surviving animals at various intervals.

7.3.6 ANALYSIS OF RESULTS

Chronic and subchronic studies rely on normal distribution of effects based on the selection of appropriate dosage ranges. Normal distribution requires a linear or logarithmic dose–response effect using corresponding dosage groups, that is, linear or logarithmic increases of dosages, respectively. Consequently, valid dose–effect responses depend on the appearance of expected effects at the highest and lowest doses. These responses are based on predicted levels of parameters based on dosage groups and include the FEL, LOAEL, LOEL, NOAEL, and NOEL. Table 7.5 describes these parameters and their significance in evaluating results of chronic and subchronic toxicology tests. Further descriptions of hazard assessment are addressed in Chapter 4.

TABLE 7.5
Monitoring Hazard Risk Assessment Based on Results of Chronic or Acute Studies

Risk Assessment Parameter	Description
AHE	**Adverse health effect:** change in body function or cell structure that might lead to disease or health problems
LOAEL	**Lowest observed adverse effect level:** EPA's lowest level of a stressor that causes statistically and biologically significant differences in test samples as compared to other samples not subjected to stressor
MRL	**Minimal risk level:** ATSDR estimate of daily human exposure to hazardous substance at or below a level that is unlikely to pose a measurable risk of harmful (adverse) non-cancerous effects
NAPHH	**No apparent public health hazard:** category used in ATSDR's public health assessments for sites where human exposure to contaminated media may occur, may have occurred in past, or may occur in future, but exposure is not expected to cause any harmful health effects
NOAEL	**No observed adverse effect level:** greatest concentration or amount of a substance, found by experiment or observation, that causes no detectable adverse alteration of morphology, functional capacity, growth, development, or life span of target organism under defined conditions of exposure
NOEL	**No observed adverse effect level:** EPA's exposure level at which no statistically or biologically significant differences in the frequency or severity of any effect in the exposed or control populations occur
PHH	**Public health hazard:** category used in ATSDR's public health assessments for sites that pose public health hazards because of long-term exposures (more than 1 yr) to sufficiently high levels of hazardous substances or radionuclides that could result in harmful health effects
RfD	**Reference dose:** EPA estimate, with built-in uncertainty or safety factors, of daily lifetime dose of substance that is unlikely to cause harm in humans
UF	**Uncertainty factor:** mathematical adjustments for reasons of safety when knowledge is incomplete; factors used in calculation of doses that are not harmful to a population; applied to LOAEL, NOAEL, MRL
UPHH	**Urgent public health hazard:** category used in ATSDR's public health assessments for sites where short-term exposures (less than 1 yr) to hazardous substances or conditions could result in harmful health effects requiring rapid intervention

Note: ATSDR = Agency for Toxic Substances & Disease Registry, part of U.S. Department of Health and Human Services, http://www.atsdr.cdc.gov; EPA = U.S. Environmental Protection Agency, http://www.epa.gov

SUGGESTED READINGS

Allaben, W.T. et al., FDA points-to-consider documents: the need for dietary control for the reduction of experimental variability within animal assays and the use of dietary restriction to achieve dietary control, *Toxicol. Pathol.*, 24, 776, 1996.

Barlow, S.M., Agricultural chemicals and endocrine-mediated chronic toxicity or carcinogenicity, *Scand. J. Work Environ. Health*, 31, 141, 2005.

Combes, R., Gaunt, I., and Balls, M., A scientific and animal welfare assessment of the OECD Health Effects Test Guidelines for the safety testing of chemicals under the European Union REACH system, *Altern. Lab. Anim.*, 34, 77, 2006.

Cooper, R.L. et al., A tiered approach to life stages testing for agricultural chemical safety assessment, *Crit. Rev. Toxicol.*, 36, 69, 2006.

Lai, M.W. et al., Annual report of the American Association of Poison Control Centers' national poisoning and exposure database, *Clin. Toxicol.*, 44, 803, 2006.

Lin, K.K., Progress report on the guidance for industry for statistical aspects of the design, analysis, and interpretation of chronic rodent carcinogenicity studies of pharmaceuticals, *J. Biopharm. Stat.*, 10, 481, 2000.

International Conference on Harmonisation, Guidance on the duration of chronic toxicity testing in animals (rodent and nonrodent toxicity testing), availability, U.S. Food and Drug Administration, *Fed. Regist.*, 64, 34259, 1999.

Van Cauteren, H. et al., The industry view on long-term toxicology testing in drug development of human pharmaceuticals, *Pharmacol. Toxicol.*, 86, 1, 2000.

REVIEW ARTICLES

Cunha, G.C. and van Ravenzwaay, B., Evaluation of mechanisms inducing thyroid toxicity and the ability of the enhanced OECD Test Guideline 407 to detect these changes, *Arch. Toxicol.*, 79, 390, 2005.

Doe, J.E., Lewis, R.W., and Botham, P.A., Comments on a scientific and animal welfare assessment of the OECD Health Effects Test Guidelines for the safety testing of chemicals under the European Union REACH system, *Altern. Lab. Anim.*, 34, 111, 2006.

Gelbke, H.P., Hofmann, A., Owens, J.W., and Freyberger A., The enhancement of the subacute repeat dose toxicity test OECD TG 407 for the detection of endocrine active chemicals: comparison with toxicity tests of longer duration, *Arch. Toxicol.*, 2006 [Epub ahead of print].

Hard, G.C. and Khan, K.N., A contemporary overview of chronic progressive nephropathy in the laboratory rat, and its significance for human risk assessment, *Toxicol. Pathol.*, 32, 171, 2004.

Hoffmann, S. and Hartung, T., Toward an evidence-based toxicology, *Hum. Exp. Toxicol.*, 25, 497, 2006.

Hutchinson, T.H., Shillabeer, N., Winter, M.J., and Pickford, D.B., Acute and chronic effects of carrier solvents in aquatic organisms: a critical review, *Aquat. Toxicol.*, 76, 69, 2006.

Kroes, R. et al., Threshold of toxicological concern for chemical substances present in the diet: a practical tool for assessing the need for toxicity testing, *Food Chem. Toxicol.*, 38, 255, 2000.

McCarty, L.S. and Borgert, C.J., Review of the toxicity of chemical mixtures: theory, policy, and regulatory practice, *Regul. Toxicol. Pharmacol.*, 45, 119, 2006.

Tetko, I.V. et al., Can we estimate the accuracy of ADME-Tox predictions? *Drug Discov. Today*, 11, 700, 2006.

8 Acute Dermal and Ocular Toxicity Testing

8.1 INTRODUCTION

United States regulatory agencies currently have testing guidelines for chemicals and materials with high potentials for accidental, inadvertent, or clinical human exposure. These substances are intended for cutaneous and ocular administration as drugs and biologicals to humans and animals and for commercial chemical (industrial, environmental, occupational, and domestic) applications.

The needs for local and systemic *in vivo* cutaneous toxicity evaluations are mandated by United States and European Union regulations. Thus, the designs of studies to evaluate acute dermal and ocular toxicities require animal models to ensure safety of product formulations before arrival on the market for human consumption. The procedures incorporated into *in vivo* studies are unique to cutaneous and ocular toxicity evaluation and the methods used for *in vivo* toxicity testing of these products are discussed below.

8.2 ACUTE DERMAL TOXICITY TESTS

8.2.1 Description

All products for human and veterinary consumption are regulated for evaluation for toxicology testing. With dermal toxicity studies, rabbits serve as standards for evaluating local toxicity (irritation). Guinea pigs are generally listed in testing guidelines as an acceptable species, but their use is uncommon for dermal toxicology evaluations. They have, until recently, been the species of choice for sensitization studies. The local lymph node assay (LLNA) performed in CBA/Ca or CBA/J mice is a refinement of the guinea pig assay. It is discussed in both the *in vivo* and *in vitro* toxicity testing sections of this book because it is an alternative to the traditional guinea pig assay.

8.2.2 Primary Irritation

Dermal acute studies are designed to provide information on local effects, particularly skin irritation and corrosion. In the aftermath of World War II and the development of chemical warfare research, the need to ensure the safety of consumer products generated significant interest. Thus the Draize skin test was born in the

TABLE 8.1
Grading of Skin Reactions: Acute Dermal Irritation

Dermal Irritation	Skin Reaction	Score
Erythema and eschar formation	No erythema	0
	Very slight erythema, barely perceptible	1
	Well defined erythema	2
	Moderate to severe erythema	3
	Severe erythema (beet redness) to eschar formation (injuries in depth)	4 (maximum)
Edema formation	No edema	0
	Very slight edema (barely perceptible)	1
	Well defined edema	2
	Moderate edema (raised ~1 mm)	3
	Severe edema (extending beyond area of exposure and raised >1 mm)	4

Source: Adapted from Draize, J.H. and Kelley, E.A., *Toxicology*, 1, 267, 1959.

mid-20th century, motivated by the hazards of unsafe cosmetics. Justified by an urgency for public protection, the Draize test became a government-endorsed method for evaluating the safety of materials meant for local contact uses. The Draize eye test is discussed in Section 8.3.

Draize et al. (1944) published a quantitative assessment of skin irritation as a guideline for product safety. They defined a primary local irritant as a substance that produced an inflammatory dermal reaction. The inflammatory process as it pertained to dermal irritation was characterized by the presence of edema* and erythema†. Table 8.1 outlines the grading of primary reactions of rabbit skin based on Draize observations. Because of the potential for pain and skin damage resulting from cutaneously applied materials, careful monitoring of cutaneous irritation and toxicity studies is now mandated by animal welfare regulations.

The Animal Welfare Act (amended 1985, U.S. Department of Agriculture) regulates the care and use of most animals used for research and toxicity testing. Although rats and mice are not covered by the act, most institutions have adopted similar internal standards for these species. The act states that skin irritancy testing is an example of a procedure that can be expected to cause more than momentary or slight pain; while the dosing procedure is generally not painful, the reaction caused by a product may elicit pain.

The rabbit is the species of choice for dermal irritation because it has relatively sensitive skin as compared to human skin. This species is also easier to handle than larger species and rabbit skin has high permeability. Because of the rabbit's enhanced sensitivity to dermal insult, however, it is generally considered to be over-predictive of human irritation, thus bringing into question its relevance to human risk assess-

---

* Accumulation of fluid in subcutaneous and interstitial spaces.
† Redness of skin as a result of increased local blood flow.

TABLE 8.2
Classification of Dermal Irritation Potential

PDII Classification	Score
Non-irritant	0.0
Negligible irritant	>0.0 to 0.5
Mild irritant	>0.5 to 2.5
Moderate irritant	>2.5 to 5.0
Severe irritant	>5.0 to 8.0

Note: Score based on observation of area at 60 min, 24, 48, and 72 hr from application. Observations for reversibility made for up to 14 days.

Source: Adapted from Auletta, C.S., *Basic Clin. Pharmacol. Toxicol.,* 95, 201, 2004.

ment. Many current regulations specify albino rats as preferable to rabbits although rabbits still represent the standard for local toxicity (irritation) evaluations. The guinea pig is generally listed in testing guidelines as an acceptable species, but is rarely used in dermal irritation toxicity evaluations.

8.2.2.1 Study Design and Procedures

Dermal irritation studies are designed to mimic human exposure and are generally performed on one to three albino rabbits. A test area is prepared by shaving the fur (2 to 3 cm^2) on the back or abdomen. Initially, a single animal is exposed sequentially (3 min and 1 and 4 hr) to the test material (0.5 ml or 0.5 g). Most drugs and biological materials are left uncovered. Applications may be covered if a material is applied under a bandage (burn treatment) or covered with clothing (to mimic occupational exposure to chemicals). A semi-occlusive covering generally consists of a porous dressing such as gauze held in place with non-irritating tape. If dermal corrosion is seen after exposure, the test is terminated and the material is classified as corrosive. If no corrosion is seen, two additional animals are exposed to the material for up to 4 hr and irritation is scored according to the Draize irritation potential classification, as detailed in Table 8.2. The sum of the mean erythema and edema scores is computed and listed as the primary dermal irritation index (PDII).

Other procedural details to be considered include (1) choice of solvent vehicle for administration, (2) amount of sample administered to the skin site, (3) need for animal restraining equipment, and (4) preparation of intact and abraded skin sites. These details are summarized in Table 8.3.

8.2.3 SKIN SENSITIZATION

Evaluation of the potential to produce sensitization is required for many chemicals. Upon application to and penetration of the skin, a chemical can elicit a subsequent reaction, usually days or weeks after repeated exposures. This type of skin sensitization

TABLE 8.3
Other Factors Involved in Dermal Irritation Study Design

Factor	Description
Solvent vehicle	Inert (CMC, saline); oil based only according to particular application; paste for solids
Sample volume	Minimized and uniform coverage to area
Restraining devices	Only in longer studies
Preparation of test site	Intact or abraded*
Post-treatment observation	Up to 4 hr; scored as described above

* Thicker of skin is gently scraped with steel comb or punctated with hypodermic needle to allow increased access to underlying subcutaneous structures.

is known as allergic contact sensitivity or dermatitis. The response requires at least 24 to 72 hr for development and a reaction is characterized as a type IV (delayed-type hypersensitivity) cell-mediated immunity. Responses range from mild irritation resulting in erythema and induration, to the development of eczema, eruptions, and flaring, to full systemic immune manifestations. The response to a chemical is typically dose-dependent and self-limiting (days to weeks) and of varying intensity. Furthermore, the process requires initial sensitization, adaptation, and subsequent re-challenge.

Immunologically, a reaction involves antigen-specific T cell activation and begins with an intradermal or mucosal challenge (*sensitization stage*). CD4$^+$ T cells then recognize MHC-II (major histocompatibility class II) antigens on antigen-presenting cells (APCs; Langerhans cells, for example) and differentiate to T_H1 cells. This sensitization stage requires prolonged (at least 2 wk) local contact with an agent. A subsequent repeat challenge stage (*elicitation*) induces differentiated T_H1 (memory) cells to release cytokines, further stimulating attraction of phagocytic monocytes and granulocytes. The release of lysosomal enzymes from the phagocytes results in local tissue necrosis. Contact hypersensitivity resulting from prolonged exposure to plant resins and jewelry, for example, is caused by the lipophilicity of a chemical in oily skin secretions, thus acting as a hapten. Although most reactions are self-limiting, extensive exposure results in tissue necrosis.

8.2.3.1 Study Design and Procedures

Invariably, the test animals used for skin sensitization studies are guinea pigs, particularly because of their known susceptibility to a variety of chemical sensitizers. Consequently, the most common skin sensitization methods include the guinea pig maximization test (GPMT) of Magnusson and Kligman and the Buehler test. Studies utilize 10 to 20 animals in a treated group and 5 to 10 in a control group. For the GPMT, induction is started according to the adjuvant method by a set of intradermal injections (test article, Freund's adjuvant, and test article in adjuvant), followed 1 wk later by topical application, and topical challenge 2 wk later.

In the closed-patch Buehler test, three topical applications 1 wk apart constitute the induction phase, followed by topical challenge 2 wk later. The appearance of edema or erythema after the challenge dose, greater than that resulting from the sensitizing dose, is indicative of sensitization. For both studies, periodic reliability checks are performed with a known mild to moderate skin sensitizer.

The murine local lymph node assay (LLNA) is a scientifically validated alternative used to assess the potentials of substances to produce skin sensitization.* The LLNA uses young adult female CBA mice instead of guinea pigs. The procedure is completed in a week and causes less pain and distress. It is based on the visual evaluation of the ability of a chemical to cause allergic reactions after repeated applications to the skins of about 20 to 30 animals. Three days after the final application, animals receive intravenous injections of ^{3}H-thymidine or ^{125}I. After 5 hr, auricular lymph nodes are excised, drained, and radioactivity is standardized according to a stimulation index (SI). A cutoff value greater than or equal to 3 is considered indicative of sensitization. Although the validity of the assay was endorsed by independent peer review evaluation (ICCVAM/NICEATM 2004; OECD Guideline 429, 2002), some concerns involve the potential for false positive results resulting from irritation and poor absorption of aqueous-based formulations.

The revised OECD Guideline 404 also suggests that data on skin irritation or corrosion can be obtained from other animal studies. For instance, it is possible to determine the irritation potential or lack of irritation potential by reviewing the results of a dermal toxicity study. In addition, if a material is highly toxic by the dermal route, it is not necessary to assess its irritation potential also. Although not specifically mentioned, it may also be possible to gain information on the irritation potential of a material from results of studies assessing dermal sensitization or the studies commonly performed to confirm skin safety in humans.

8.2.4 PHOTOALLERGIC AND PHOTOTOXIC REACTIONS

Dermal phototoxicity of a chemical is defined as a toxic response elicited after dermal or systemic exposure to a chemical following subsequent exposure to light, particularly ultraviolet (UV) radiation from sunlight. The biological effects of UV rays alone are of lower frequency and longer wavelength (about 10^{-8} to 10^{-6} m) than ionizing radiation (10^{-7} to 10^{-12} m for x-rays). Thus the effects of UV radiation are less penetrating and benign.

Unlike ionizing radiation, skin damage induced by UV rays is mediated principally by the generation of reactive oxygen species (ROS) and the interruption of melanin production. Like ionizing radiation, however, cumulative or intense exposure to UV rays precipitates DNA mutations: base pair insertions, deletions, single-strand breaks, and DNA-protein cross-links. DNA repair mechanisms play an important role in correcting UV-induced DNA damage and in preventing further consequences of excessive UV exposure (as with sunburn).

Melanin production by melanocytes increases and the epidermis thickens in an attempt to prevent future damaging effects. The protective ability of antioxidant enzymes and DNA repair pathways diminishes with age, thus setting the conditions

---

* Because it is an *in vivo* assay, the OECD considers it of equal merit but not a required replacement for animal toxicity tests.

TABLE 8.4
Chronic Effects of Sunlight and Photosensitivity Reactions

Reaction to Excessive Sunlight	Clinical Effects
General dermatologic reactions	Dermatoheliosis: aging of skin due to chronic exposure to sunlight; elastosis: yellow discoloration of skin with accompanying small nodules; wrinkling, hyperpigmentation, atrophy, dermatitis
Actinic keratoses	Precancerous keratotic lesions after many years of exposure to UV rays
Squamous/basal cell carcinomas	Occur more commonly in light-skinned individuals exposed to extensive UV rays (sun tanning) during adolescence
Malignant melanomas	Cancerous dermal lesions associated with increased, intense, prolonged exposure to UV light
Photosensitive reactions	Erythema and erythema multiform lesions; urticaria, dermatitis, bullae; thickened, scaling patches

for development of skin neoplasms later in life. In addition, UV light can convert chemicals into ROS that elicit local cellular toxicity.

The extent of solar injury depends on the type of UV radiation, the duration and intensity of exposure, clothing, season, altitude and latitude, and the amount of melanin pigment present in the skin. In fact, individuals differ greatly in their responses to sun exposure. Also, harmful effects of UV rays are filtered out by glass, smog, and smoke and enhanced when reflected off snow and sand. The indiscriminate use of chlorofluorocarbons in aerosol propellants depletes the UV-blocking properties of ozone in the stratosphere, thus allowing greater intensities of UV rays to penetrate through the protective upper atmospheric layers.

Table 8.4 summarizes the chronic effects of sunlight and photosensitivity reactions. Actinic keratoses, described as precancerous keratotic lesions, result from years of sun exposure, especially in fair-skinned individuals. They present as pink, poorly marginated, scaly, or crusted superficial growths on skin.

Photosensitive reactions involve the development of unusual responses to sunlight, sometimes exaggerated in the presence of a variety of chemicals or precipitating circumstances. Erythema (redness) and dermatitis (inflammation of the skin or mucous membranes) are acute responses to direct UV light-induced photoallergic reactions. They are mediated by dilation and congestion of superficial capillaries and a type IV hypersensitivity cell-mediated reaction. In addition, macules, papules, nodules, and target (bull's eye shaped) lesions are seen in multiform variations of erythematous reactions.

Urticaria is a pruritic skin eruption characterized by transient wheals of varying shapes and sizes, with well defined erythematous margins and pale centers. Development of urticaria is mediated by capillary dilation in the dermis in response to release of vasoactive amines (histamine and kinins) upon sun exposure, especially subsequent to chemical exposure. More severe photosensitivity reactions lead to the development of bullae — thin walled blisters on the skin or mucous membranes larger than 1 cm in diameter and containing clear serous fluid. Dehydration, scaling, scarring, fibrosis, and necrosis develop as the exposed areas heal.

Among the exogenous factors that contribute to the development of photosensitivity reactions are ingestion or application of antibacterial and antifungal antibiotics (sulfonamides, tetracyclines; griseofulvin, respectively) and thiazide diuretics. Contact with dermal products containing coal tar, salicylic acid, plant derivatives, and ingredients of colognes, perfumes, cosmetics and soaps are additional sources of the problem.

8.2.4.1 Study Design and Procedures

In vivo tests for photosensitivity and phototoxicity are conducted in guinea pigs or rabbits. Test articles are administered orally or parenterally for 10 to 14 days or applied topically. The challenge phase starts 2 to 3 wk later with another dose plus exposure to a UV lamp.* Factors that affect photosensitivity test results include a chemical's propensity to absorb light, length of UV exposure, and proximity of light source to shaved area. As with other dermal tests, control groups incorporate positive and negative photoallergic agents. Lesions are scored as described above.

8.3 ACUTE OCULAR TOXICITY TESTS

No issue in toxicology testing has provoked more controversy or protest than the Draize eye test. Published in the 1940s as a quantitative analysis of the effects of chemicals on rabbit eyes, it followed a similar manuscript by Friedenwald et al. (1944) that proposed a quantitative method for evaluating the severities of ocular testing methods. Today, advances in ocular toxicology are challenging the validity, precision, relevance, and need of the Draize eye test.

Preclinical product safety tests with rabbits and other mammals also raise ethical concerns of animal welfare. The Draize eye test, however, still remains the only official, regulatory-approved procedure for predicting ocular irritants. With the development of alternative non-animal procedures to replace the Draize test, the data generated in the Draize procedure are also used as standards against which the performances of alternative methods are measured (see Chapter 13).

8.3.1 Eye Irritation and Corrosion Testing

The Draize test involves a standardized protocol for instilling agents onto the corneas and conjunctivae of laboratory animals, particularly rabbits. A sum of ordinal-scale items of the outer eye gives an index of ocular morbidity. The test uses a scoring system that describes reactions produced in the cornea, conjunctiva, and iris. Since its inception, the test has undergone several modifications, including the addition of other descriptive parameters such as thickness of the eyelids, induction of erythema,

---

* The shorter wavelengths of sunburn-producing UVB radiation (280 to 320 nm λ) are more damaging than UVA rays (320 to 400 nm λ).

edema, discharge, corneal opacity, capillary damage, and pannus (vascularization) of the cornea.

In an attempt to standardize ocular testing, OECD Guideline 405* recommends that all existing information on a test substance be evaluated prior to undertaking *in vivo* eye irritation and corrosion studies of that substance and that sequential testing strategies be considered for each new and revised test guideline. Movement toward a weight-of-evidence approach to hazard assessment including multi-step, tiered testing strategies is suggested prior to commencing animal tests. Furthermore, since *in vivo* eye irritancy and corrosion tests are federally regulated, the revised guidelines recommend performing tests sequentially, one animal at a time. This allows reassessment of data as the results accumulate to avoid duplication of animal tests. In addition, corrosive classification of a test material precludes further testing with additional animals and allows for immediate euthanasia of injured or suffering animals. Additional experimentation is necessary only if data are insufficient to support classification.

8.3.2 OBJECTIONS TO AND LIMITATIONS OF EYE IRRITATION AND CORROSION TESTING

Several major sources of variability in animal eye irritation and corrosion testing have led to the conclusion that the rabbit eye is an unsatisfactory model for the human eye. The individual testing of animals creates small group sizes for multiple dose testing, thus necessarily disqualifying adequate statistical comparisons. Increasing group size, however, would undermine the impetus toward reduction of animal testing and development of *in vitro* alternatives to the Draize eye method. Also, grade definitions of ocular irritancy and corrosion are relatively oversimplified. Thus the existing scoring systems do not reflect the complexities of the total *in vivo* response. Also, the use of algorithms to simplify the *in vivo* data is not sufficiently accurate for comparison with *in vitro* data. These and other considerations that add to the variability and objections to the use of the Draize eye test are defined in Table 8.5.

8.4 CONCLUSIONS

In vivo skin and eye toxicology testing methods have been used for decades to provide data for formulating risk assessment programs. The information generated during these tests is hampered by the scientific limitations and ethical objections posed, with the animal care issue at the forefront. The development of *in vitro* alternatives for this and other toxicity tests (discussed in Part III) is necessarily based on comparisons to *in vivo* data. Thus validation of alternative methods depends on the reliability of *in vivo* data. It is logical to conclude that until the issues concerning animal skin and eye testing methods are addressed, development of alternative *in vitro* tests will be more difficult than initially assumed.

---

* *Acute eye irritation/corrosion* was adopted in 1981 and revised in 1987 and 2002.

TABLE 8.5
Sources of Variability and Objections to Draize Eye Irritation and Corrosion Testing

Source	Clinical Effects and Objections
Laboratory variability	Intra- and inter-laboratory variations observed during subjective scoring of test substances; standardization of time-to-heal criteria for assessment of severity of damage; grade definitions oversimplified; existing scoring systems do not reflect complexities of *in vivo* responses; lack of consistent data collection and interpretation of inadequate algorithms
Animal-related variability	Variations in individual responses of animals exposed to same test substance; significant differences in rabbit versus human eye anatomy: presence of nictitating membrane and lower mean corneal thickness in rabbit eye; differences in tearing mechanisms; increased sensitivity of rabbit (versus human or primate) eye to chemical irritants; sensory response at time of instillation
Standardization of experimental variability	Concentration of test substance; use of anesthesia altering ocular responses; irrigation procedures to remove excess or particulates; adoption of standard dose volume (10 µl) to improve reproducibility; small numbers of animals used per group; duration of exposure; time to heal after exposure
Ethical considerations	Minimizing acute and chronic pain and suffering; proper implementation of euthanasia

SUGGESTED READINGS

Basketter, D.A. et al., Threshold for classification as a skin sensitizer in the local lymph node assay: a statistical evaluation, *Food Chem. Toxicol.*, 37, 1, 1999.

Draize, J.H. and Kelley, E.A., The urinary excretion of boric acid preparations following oral administration and topical applications to intact and damaged skin of rabbits, *Toxicology*, 1, 267, 1959.

Draize, J. H., Woodward, G., and Calvery, H.O., Methods for the study of irritation and toxicity of substances applied topically to the skin and mucous membranes, *J. Pharmacol. Exp. Ther.*, 82, 377, 1944.

Friedenwald, J.S. and Buschke, W., Some factors concerned in the mitotic and wound-healing activities of the corneal epithelium, *Trans. Am. Ophthalmol. Soc.*, 42, 371, 1944.

ICCVAM, The murine local lymph node assay, NIH Publication 99-4494, Research Triangle Park, NC, 1999, http://iccvam.niehs.nih.gov/methods/epiddocs/cwgfinal/cwgfinal.html.

Jester, J.V., Extent of corneal injury as a biomarker for hazard assessment and the development of alternative models to the Draize rabbit eye test, *Cutan. Ocul. Toxicol.*, 25, 41, 2006.

Magnusson, B. and Kligman, A.M., The identification of contact allergens by animal assay: the guinea pig maximization test, *J. Invest. Dermatol.*, 52, 268, 1969.

Martinez, V. et al., Evaluation of eye and skin irritation of arginine-derivative surfactants using different *in vitro* endpoints as alternatives to *in vivo* assays, *Toxicol. Lett.*, 164, 259, 2006.

Secchi, A. and Deligianni, V., Ocular toxicology: the Draize eye test, *Curr. Opin. Allergy Clin. Immunol.*, 6, 367, 2006.

Wilhelmus, K.R., The Draize eye test. *Surv. Ophthalmol.*, 45, 493, 2001.

REVIEW ARTICLES

Abraham, M.H. et al., Draize rabbit eye test compatibility with eye irritation thresholds in humans: a quantitative structure-activity relationship analysis, *Toxicol. Sci.,* 76, 384, 2003.

Auletta, C.S., Current *in vivo* assays for cutaneous toxicity: local and systemic toxicity testing, *Basic Clin. Pharmacol. Toxicol.,* 95, 201, 2004.

Basketter D.A. et al., Evaluation of the skin sensitizing potency of chemicals by using the existing methods and considerations of relevance for elicitation, *Cont. Derm.,* 52, 39, 2005.

Buehler, E.V., Occlusive patch method for skin sensitization in guinea pigs: the Buehler method, *Food Chem. Toxicol.,* 32, 97, 1994.

Chatterjee, A., Babu, R.J., Ahaghotu, E. and Singh, M., The effect of occlusive and unocclusive exposure to xylene and benzene on skin irritation and molecular responses in hairless rats, *Arch. Toxicol.,* 79, 294, 2005.

Frankild, S., Volund, A., Wahlberg, J.E., and Andersen, K.E., Comparison of the sensitivities of the Buehler test and the guinea pig maximization test for predictive testing of contact allergy, *Acta Derm. Venereol.,* 80, 256, 2000.

Kimber, I. et al., Skin sensitization testing in potency and risk assessment, *Toxicol. Sci.,* 59, 198, 2001.

Maurer, J.K., Parker, R.D., and Carr, G.J., Ocular irritation: microscopic changes occurring over time in the rat with surfactants of known irritancy, *Toxicol. Pathol.,* 26, 217, 1998.

McDougal, J.N. and Boeniger, M.F., Methods for assessing risks of dermal exposures in the workplace, *Crit. Rev. Toxicol.,* 32, 291, 2002.

OECD, Guideline 429: Murine Local Lymph Node Assay, http://www.oecd.org.

Prinsen, M.K., The Draize eye test and *in vitro* alternatives: a left-handed marriage? *Toxicol. in Vitro,* 20, 78, 2006.

Roberts, J.E., Screening for ocular phototoxicity, *Int. J. Toxicol.,* 21, 491, 2002.

Robinson, M.K., Nusair, T.L., Fletcher, E.R., and Ritz, H.L., A review of the Buehler guinea pig skin sensitization test and its use in a risk assessment process for human skin sensitization, *Toxicology,* 61, 91, 1990.

Robinson, M.K. and Perkins, M.A., A strategy for skin irritation testing, *Am. J. Contact Derm.,* 13, 21, 2002.

Svendsen, O., The minipig in toxicology, *Exp. Toxicol. Pathol.,* 57, 335, 2006.

York, M. and Steiling, W., A critical review of the assessment of eye irritation potential using the Draize rabbit eye test, *J. Appl. Toxicol.,* 18, 233, 1998.

9 Toxicity Testing for Fertility and Reproduction

9.1 INTRODUCTION

Teratogens are chemicals or drugs that cause toxicity to an embryo by inducing structural malformations, physical dysfunction, behavioral alterations, or genetic abnormalities in the fetus. The expression of the teratogenic response — the embryotoxic effect of a compound on the growth and development of the fetus — is usually manifested at birth or in the immediate post-natal period. A chemical may also interfere with reproduction by inducing toxicity early in the reproductive cycle (first trimester) by impairing fertilization or by interfering with implantation. Thus, the initial interaction of a fertilized ovum and exposure to a suspected teratogen may result in the inability to establish pregnancy or induction of its spontaneous termination.

Reproductive toxicology also involves the study of adverse effects of xenobiotic agents on events prior to fertilization of the ovum including spermatogenesis and oogenesis. These responses created a need to define, develop, and properly perform tests aimed at evaluating chemicals with the potentials to interfere with human reproduction.

Therapeutic drug use and exposure to environmental and occupational agents during pregnancy have increased dramatically in recent years. Epidemiological studies indicate that about 90% of women ingest one or more medications while pregnant (FDA, 1995). In addition, the rates of environmental and occupational exposure to hazardous chemicals before or during pregnancy have increased substantially. Excluding the commonly prescribed prenatal vitamins, iron supplements, and tocolytic* drugs, women under 35 years of age take three prescriptions during the course of their pregnancies on average; the number increases to five in women over 35.

Some of the environmental chemicals to which women are inadvertently exposed during pregnancy include insecticides, herbicides, fungicides, metals (such as mercury in seafood), and growth hormones (in processed foods). For most of these chemicals and drugs, well controlled studies were not conducted prior to marketing nor are the studies mandated for most therapeutic or non-therapeutic agents.

---

* Used to suppress premature labor.

One complicating factor arises from the extensive abilities of therapeutic drugs, environmental chemicals, and herbal products to infiltrate the maternal–fetal environment. Although most therapeutic drugs are screened for their embryotoxic or teratogenic potentials in animal reproductive studies, many chemicals are not. United States regulations do not mandate that pharmaceutical and chemical companies provide such information. Also, most drug development programs do not routinely incorporate screening or follow-up studies of drugs for potential embryotoxic effects in pregnant women during preclinical and clinical phases or in post-marketing periods. Finally, environmental chemicals have generally not been subjected to teratogenic testing. This means no human embryotoxic data exist for a multitude of chemicals used in the marketplace.

9.2 HISTORY AND DEVELOPMENT OF TERATOGENICITY TESTING

The thalidomide disaster of 1961 initiated a significant effort by United States and European legislative bodies to establish programs for teratogenicity testing. Developed as a sedative–hypnotic with no particular advantage over drugs of the same class, thalidomide was initially shown to lack teratogenic effects in all species tested except rabbits. Soon after its introduction to the European market, it was linked to the development of a relatively rare birth defect known as *phocomelia*. The epidemic proportions of the teratogenic effect of thalidomide prompted the passage of the Harris-Kefauver Amendment in the United States in 1962, one of the many additions to the Pure Food and Drug Act. The amendment requires extensive preclinical pharmacological and toxicological research before a therapeutic compound is marketed.

9.3 BRIEF DESCRIPTION OF MATERNAL–FETAL PHYSIOLOGY

9.3.1 FETAL–PRENATAL DEVELOPMENT

Prenatal growth and development involves the beginning, growth, and maturity of the anatomy and physiology from fertilization (conception) to parturition (birth). The first trimester (first 10 wk of pregnancy in humans) is characterized as the period for *embryological development* during which differentiation of precursor stem cells progresses, culminating in the appearance of the fetal membranes and embryonic disk. The remainder of the pregnancy (11 to 40 weeks) is dedicated to *fetal development* during which the established blueprints of organs and tissues undergo further growth and maturity. These processes constitute prenatal growth and development. Thus, in humans, the duration of the gestational period is normally 9 months (or 280 days from the first day of the last menstrual period, assuming a regular 28-day cycle). *Postnatal development* begins at birth and progresses to adolescence.

Figure 9.1 illustrates the cycle of reproductive life. The major stages of reproduction (shaded areas) are composed of the events that occur during formation and development of the embryo within those stages (darker shaded area), along with the parallel time sequences (gray boxes). Consequently the most sensitive periods of

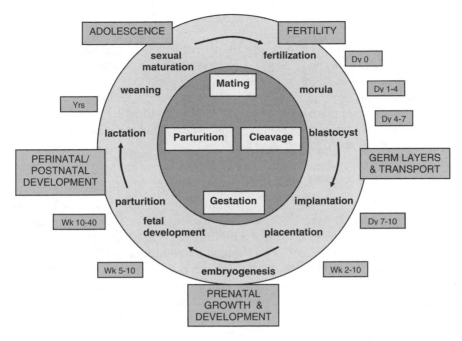

FIGURE 9.1 Cycle of reproductive life and periods of potential embryo and reproductive toxicity.

exposure to toxic substances correspond to the events occurring during the reproductive cycles (mating, cleavage, gestation, parturition). In general, exposure during the earlier stages of fetal development is associated with more severe teratogenic consequences. In addition, dose and duration of exposure are critical in evaluating the mechanisms of reproductive toxicity.

9.3.2 FIRST TRIMESTER

The first trimester is the period corresponding to embryological and early fetal development. The main events occurring during the first trimester include:

Cleavage (days 1 through 6) — A series of cell divisions following fertilization of the ovum by the sperm cell, resulting in the formation and transport of the morula (early cell division stage) and blastocyst (late cell division stage) through the fallopian tube.

Implantation (days 7 through 10) — Adhesion, attachment, and penetration of the uterine endometrial lining by the blastocyst; trophoblast, amniotic cavity, and inner cell mass form and expand. Further development of the yolk sac, chorion, allantois, amnion, and embryonic disc (from the inner cell mass) ensues during weeks 3 and 4 (beginning of embryogenesis).

Placentation (weeks 2 through 10) — Starting as early as day 10, blood vessels form around the periphery of the blastocyst to establish the presence of the placenta; this multilayered membranous structure is composed of infiltrating chorionic villi and highlighted by the appearance of fetal blood vessels.

Embryogenesis (weeks 5 through 10, concurrent with placentation) — The differentiation and folding of the embryonic disc produces a physically and developmentally distinct and recognizable embryo. Because of the critical events occurring during this phase, as well as during fertilization and formation of germ layers (Figure 9.1), the first trimester represents the most sensitive period of prenatal growth and development to external stimuli, especially to drugs and chemicals.

9.3.3 SECOND AND THIRD TRIMESTERS

The second and third trimesters range from the 10th to 40th weeks (4th to 9th months) of gestation. The second trimester is characterized by fetal development of organ systems (organogenesis) while the third trimester exhibits rapid growth and maturation of established organs in preparation for parturition (birth).

The next section summarizes the criteria necessary to develop well planned reproductive toxicity tests according to the mammalian species under consideration. Extrapolation of the data derived from animal studies to human hazard assessment is also discussed. The current focus on the development of *in vitro* alternative toxicity tests for reproductive studies is covered in Part III, Chapter 16.

9.4 MECHANISMS OF DEVELOPMENTAL TOXICITY

9.4.1 SUSCEPTIBILITY

As noted above, the first 8 to 10 weeks of embryogenesis qualify as a sensitive time period for the exposure of a fetus to drugs or chemicals. For instance, exposure to exogenous compounds within 21 days after fertilization presents the greatest potential for embryotoxicity resulting in the death of the fetus. Exposure to compounds between the 3rd and 8th weeks is potentially teratogenic, that is, the agents may interfere with organogenesis. The effects are distinguished by the development of gross anatomic, metabolic, or functional defects and precipitation of spontaneous abortion.

Because the schematic for embryonic development is already established by the beginning of the second trimester, exposure to toxic agents during the second or third trimesters is less likely to be teratogenic. However, exposure still may interfere with growth and maturation of critical organ systems, resulting in underdevelopment or alterations of normal organ function.

9.4.2 DOSE–RESPONSE AND THRESHOLD

Exchange of nutrients, gases, and waste materials between the maternal and fetal circulations commences from the 5th week after fertilization, with the initiation of the formation of the placenta. Maternal and fetal circulations do not mix. Instead, maternal blood enters intervillous spaces (sinuses) of the placenta through ruptured maternal arteries, then drains into uterine veins for return to the maternal circulation.

Fetal blood enters the placenta via a pair of umbilical arteries and returns to the fetal circulation by the umbilical vein. The umbilical arteries branch into capillaries, are surrounded by the syncytial trophoblasts, and form a network of chorionic villi. Solutes, gases, and nutrients from the maternal circulation enter the sinuses, surround and bathe chorionic villi, and traverse the epithelium and connective tissues of the villi before penetrating the fetal capillary endothelial cells. The materials are then carried toward the embryo through the umbilical vein.

Chemicals enter the fetal circulation as they traverse any other membrane barrier. The rate of exchange between maternal and fetal circulation depends on the established equilibrium across the placenta and the rate of penetration of the chemical through the endothelial and epithelial barriers. Lipid-soluble drugs and chemicals and dissolved gases readily infiltrate placental membranes and establish high concentrations of the substances in venous cord blood. Thus, compounds may freely circulate around the embryo within 30 to 60 min of maternal administration. In addition, exogenous substances can alter dynamic tone of placental blood vessels, altering nutrient, gas, and waste exchange. Depending on the toxicokinetics, some chemicals such as gases restrict oxygen transport to fetal tissues, change biochemical dynamics of the two circulations, and interfere with the endocrine function of the placenta. The degree of chemical penetration of placental barriers relies less on the number and thickness of the layers and more on the dynamic physicochemical interactions of the chemical and placenta.

It is important to consider that because embryonic tissues grow rapidly and have high DNA turnover rates, they are vulnerable to many agents that may not promote toxic reactions in children or adults. As explained above, the most sensitive period for embryonic development is within the first trimester of pregnancy, when the fetus is in the early stages of embryonic development. Some classes of agents require repeated, continuous, or high dose exposure to elicit untoward reactions. Other more potent chemicals such as suspected carcinogens target rapidly proliferating cells and tissues. Their effects are significant and occur with higher frequency.

9.5 TERATOGENICITY

9.5.1 TERATOGENIC AGENTS

According to the definition of a teratogen cited at the beginning of this chapter, a teratogenic response to a chemical involves an embryotoxic effect on the growth and development of a fetus. The effect is manifested at birth or in the immediate post-natal period. Approximately 25% of all teratogenic responses are due to genetic susceptibilities and chromosomal aberrations while less than 75% arise from unknown origins. The remaining, apparently small, percentage of teratogenic responses arises from known environmental and therapeutic exposures and sociobehavioral habits. Consequently, these recognized human teratogens are subject to intensive scrutiny as a result of their continuous presence and important uses in society. The American College of Occupational and Environmental Medicine recognizes several classes of known human teratogens; they are listed in Table 9.1.

TABLE 9.1
Classes of Teratogens Recognized by American College of Occupational and Environmental Medicine

Classes of Known Human Teratogens	Examples
Ionizing radiation	Atomic weapons, radioiodine, radiation therapy
Maternal infections	Cytomegalovirus, herpes viruses I and II, parvovirus B-19, rubella virus, syphilis, toxoplasmosis, Venezuelan equine encephalitis virus
Maternal disorders	Alcoholism, endemic cretinism, diabetes, folic acid deficiency, hyperthermia, phenylketonuria, rheumatic disease and congenital heart block, virilizing tumors
Additive behavioral habits	Chronic alcoholism, drug abuse, cigarette smoking
Environmental chemicals	Metals (including mercury and lead compounds), herbicides, industrial solvents, polychlorobiphenyls (PCBs)
Drugs	Many therapeutic agents

9.5.2 FACTORS ASSOCIATED WITH TERATOGENICITY

Any chemical agent should be suspected of causing a teratogenic response if the conditions of exposure are sufficient for eliciting a toxic reaction. For instance, *dose–response* and *time of exposure* to a particular agent often determine the severity of the damage and the type of defect that occurs. In general, the teratogenic effect of a chemical is directly proportional to the toxic response. Consequently, a relatively higher dose of a chemical is more likely to trigger a teratogenic effect than a lower dose. Time of exposure is another critical component because vulnerability to chemical exposure is greater early in embryonic development (first trimester) than during the fetal development period (second and third trimesters).

The experience with thalidomide provides a classic example of critical exposure during organogenesis. The specificity of the malformations was linked to the time of exposure to the sedative–hypnotic during pregnancy (after the last menstrual period): 35 to 37 days, ears; 39 to 41 days, arms; 41 to 43 days, uterus; 45 to 47 days, tibia; 47 to 49 days, triphalangeal thumbs.

The types and severity of abnormalities caused by a teratogenic agent are also more dependent on the *genetic susceptibility* of a pregnant woman than on the susceptibility of the fetus. For example, variation in maternal metabolism toward a particular drug determines what metabolites eventually reach the fetus. Differences in placental membranes, rate of formation, placental transport, and biotransformation also affect fetal exposure.

Finally, as noted above, the principles of toxicokinetics and pharmacokinetics of different classes of compounds influence the degree of penetration of the placental membrane. Generally, the greater the lipid solubility of an agent, the more likely is the potential for fetal exposure through the maternal–fetal circulation.

Thus, these factors as well as the characteristics of individual compounds determine the probability that a chemical will cause a genetic defect. The probabilities are categorized accordingly:

Absolute risk — The rate of occurrence of an abnormal phenotype among individuals exposed to an agent. Chronic maternal alcoholism increases the risk of an infant born with fetal alcohol syndrome up to 45%;

Relative risk — The ratio of the rate of the condition among those exposed to a chemical and the non-exposed general population. Cigarette smokers have twice the relative risk of the general population. Thus the risk of delivering a low birth weight infant is twice as likely in smokers than nonsmokers. It is important to note, however, that high relative risk may still indicate low absolute risk in rare conditions such that a high relative risk of 10 for a rare condition with an incidence of 1 in 100,000 calculates to a low absolute risk of 1 in 10,000;

Attributable risk — The rate of a condition attributed to exposure to an agent, especially to known teratogens (90% risk of development of phocomelia in patients exposed to thalidomide *in utero*).

The increasing awareness of teratogenic potential from exposure to hazardous chemicals and therapeutic drugs during pregnancy has facilitated dissemination of information among regulatory agencies, health care professionals, and the public. As a result, teratogen information services are easily accessed in most states through national databases and the efforts of applicable societies and organizations (see Suggested Readings and Review Articles cited at the end of this chapter).

9.5.3 Animal Tests for Teratogenicity

Developmental toxicity testing is primarily intended to provide hazard identification concerning the potential effects of prenatal exposure on a developing human or animal organism. For human pharmaceuticals, this information can be used to limit patient population exposure. In addition to hazard identification, in certain circumstances the no observed effect level (NOEL) determined by developmental toxicity studies for veterinary pharmaceutical compounds can be used as the basis for establishment of acceptable daily intake (ADI).

In order to mimic a suspected or known human teratogenic response, the test system involves the administration of a specific chemical encompassing a range of doses suspected of eliciting toxic effects on an animal species. Ideally, the species should be genetically susceptible to a variety of chemical classes and the animals should be at a sensitive stage of embryonic development. The degree of susceptibility, however, is less associated with a genetically programmed species for chemical sensitivity and more a function of factors associated with properties of biotransformation peculiar to an animal group. Consequently, maternal and/or fetal ability for biotransformation of compounds may alter the net effect of a teratogenic substance. Thus, depending on the parent chemical, active biotransformation reactions may predispose an embryo for teratogenic effects by forming reactive intermediates.

Alternatively, although the absence of rapid biotransformation reactions may reduce the production of active metabolites, the lack of inactivating mechanisms may prolong exposure of the fetus to the active parent compound.

The most common approach to identifying teratogens is to perform developmental toxicity tests in two species, usually rodents and rabbits, for evaluation of human therapeutic drugs [International Conference on Harmonization (ICH), 1994] and for testing of pesticides and environmental chemicals [U.S. Environmental Protection Agency (EPA), 1998]. The basis for the two-species approach is that no single species is definitively predictive for human teratogenicity nor is any single species more sensitive to teratogens. Rodents and rabbits are particularly useful because:

1. They appear to demonstrate low incidences of spontaneous congenital anomalies.
2. Care and costs are reasonable.
3. They produce large litters.
4. They have relatively short gestation times (several weeks).

For instance, rodent and rabbit gestation times are limited to 21 days and 32 days, respectively. Thus, in order to detect potential teratogenicity of a chemical, the agent is administered within 7 to 10 days (for rodents and rabbits, respectively) after implantation. This period of cell cleavage and embryogenesis represents a narrow window for detecting teratogens and developmental and reproductive toxicity in these species.

The use of a tiered approach was recently suggested for developmental toxicity evaluation of veterinary drugs for food-producing animals, that may also be useful for a variety of chemical and pharmacological classes (ICH, 1994; EPA, 1998). The tiered method begins with evaluation and observation for developmental toxicity in a rodent species. If the testing proves positive, no further testing is required in a second species (unless the ADI is determined based on the NOEL from this study). A negative or equivocal result for teratogenicity in a rodent prompts a developmental test in a second species, customarily the rabbit. Other signs of developmental toxicity in the rodent species in the absence of teratogenicity — visceral or skeletal malformations, fetal mortality, or intrauterine growth retardation — requires a developmental toxicity test in a second species. Thus, the tiered approach provides thorough hazard identification based on the use of a second species for compounds negative for teratogenicity in a rodent. This scheme for stepwise testing ensures adequate screening of potential teratogens and attempts to limit unnecessary animal testing.

9.5.3.1 Standard Protocols

Standard methodologies for screening for teratogenicity are performed to ensure sufficient exposure at critical times post-implantation. The protocol is outlined in Table 9.2. Essentially, administration of a chemical is proportioned about 6 days post-implantation to determine teratogenic potential. Dose–response relationships are based on observable and measured defects according to treatment group.

TABLE 9.2
Standard Methodology for Teratogenicity Testing in Animals

Time Period	Protocol
Day 0	Breeding of animals; determination of successful impregnation by presence of vaginal plug
Days 6 to 15: rodent; days 6 to 18: rabbit	Chemical administration to females (3 dosage groups, 10 rats or 5 rabbits per group)
~ Day 20: rodent; ~ day 30 to 31: rabbit	Euthanize animals, necropsy dams and fetuses; identify gross anomalies; perform pathological, histological, physiological, and biochemical analyses of fetuses

9.6 EMBRYOTOXICITY

As with teratogenic studies, classical methods for testing embryotoxic substances involve rodents and rabbits, principally because of their relatively short gestational periods. The procedures involve exposure of the female either before or immediately after conception for the length of gestation. The beginning of the exposure sequence depends on the desired toxicity data. For instance, to determine the effects of a chemical on growth and development of the fetus, the exposure period begins several days after fertilization, at about implantation. The time sequence must be carefully determined so as not to interfere with mating.

Rodents and rabbits are convenient for embryotoxic experiments because of the large sizes of their litters, usually about 8 to 10 embryos per litter, thus allowing for chemical scoring according to the number of fetuses affected. For instance, a particular indicator is selected as a measure of toxicity and the endpoint is then monitored in each embryo. The minimal accepted value for this indicator is regarded as toxic and the experiment is completed on the rest of the litter. The greater the number of embryos that exhibit toxicity by this criteria, the higher the score and the more toxic the chemical. For a given number of chemicals that can be screened at different dosage levels for various indicators in each fetus, the number of experiments is planned so as to optimize resources and effort. The following section outlines the various protocols available for embryotoxicity testing.

9.6.1 WHOLE EMBRYO CULTURE

9.6.1.1 Pre-Implantation Techniques

Since the first description of the culture of pre-implantation embryos was reported in 1913, the method has since shown its importance in experimental embryology and reproductive biology. The technique is described for the pre-implantation cultures of embryos from mice, rabbits, and humans. In general, 3- to 6-day old embryos (3-day morulae to 6-day blastocysts) are removed from super-ovulated sexually mature animals and transplanted into culture dishes containing media supplemented with fetal bovine serum or serum-free conditioned medium.

The cultures are exposed to the test substance during the incubation period and examined microscopically or monitored for growth and proliferation using biochemical procedures. A variety of cell viability indicators are then used to evaluate the cytotoxicities of xenobiotics in cultured embryonic stem cells derived from the mouse blastocysts. Thus, pre-implantation embryo cultures may provide interesting stem cell systems for testing embryonic toxicity after exposure to xenobiotics.

9.6.1.2 Post-Implantation Embryo Culture

A classical method for embryotoxic potential that requires separation of whole embryos from females is known as post-implantation embryo culture. The technique involves the surgical removal of 1- or 2-week old rodent embryos, after which they are allowed to acclimate in medium under cell culture conditions for several hours. The viable fetuses are then exposed to different doses of chemicals and the effects on growth and development are determined.

The method is dramatic because structural changes occur quickly, within several days, and embryotoxicity is monitored in culture as growth and differentiation progress. This method has the additional benefit of allowing for further investigation of substances with teratogenic activities as well. In addition, culturing embryos *in vitro* provides the opportunity to study the direct actions of suspected toxins free of maternal hormonal and pharmacokinetic influences. Dose-related growth retardation and frequency of malformations are monitored macroscopically and depend on the periods of gestation of the explanted embryos.

Post-implantation whole embryo culture is suitable for screening embryotoxic substances because it involves the period of fetal development that includes morphogenesis and early organogenesis. The procedure is adequate for screening chemicals that also have teratogenic potential as well as embryotoxicity because it monitors multiple cellular processes that reflect the complex nature of embryonic development. The period for identifying toxicity, however, is limited up to week 2 because development of cultured rodent embryos requires placental development. Consequently, the technique reflects activities of chemicals during several coordinated cellular events, especially proliferation, migration, association, differentiation, and cell death. Ideally, proper performance of the post-implantation embryo culture screening technique involves:

1. A sufficient number of viable fetuses explanted for testing
2. Unobstructed, progressive, and continuous embryonic development
3. Knowledge of the responses and mechanisms of chemical toxicity
4. Ease of manipulation and interpretation of the protocol and a low frequency of false negatives
5. Administration of approximate equimolar concentrations as those encountered in therapeutic or environmental exposures, with systematic control of dose and time of exposure
6. Elimination of maternal factors that may interfere with the test system's response to the agent

Occasionally, responsiveness of post-implantation rodent embryos to cytotoxic agents may require metabolic activation. Thus modifications of the protocols may require incubation of embryos in medium containing the test agent plus hepatic S9 fractions.

Although the procedure uses growth in culture as a component of the protocol, it should not be confused with *in vitro* cell culture methods (Chapter 16). The embryo cultures still require extensive animal preparation and sacrifice and offer few advantages for large scale prescreening studies.

9.6.1.3 Organ Cultures

Fetal organs are maintained in culture in the primary cell stage by establishing whole organ cultures and partially or totally submerging the tissue in growth culture medium. Briefly, the fetal tissue explant is placed on micron filter inserts or in tissue culture plastic coated with extracellular matrix components. This approach has been useful in the culture of limb buds from several species of animals in a serum-containing or chemically defined medium. Morphological development and differentiation are monitored microscopically; biochemical indicators are also employed.

Several organs have been extensively investigated for embryotoxic phenomena including kidney development from suspected nephrotoxic agents and primary organotypic cultures of type II epithelial cells from fetal rat lungs. Thus, as with other organ-specific primary cultures, the advantage of organ culture lies in the ability of the *in vivo–in vitro* system to act as a model for detecting target organ toxicity.

9.6.1.4 International Guidelines

In 1999, the Organization for Economic Cooperation and Development (OECD) proposed international recommendations as harmonization guidelines of chemical classification systems for reproductive toxicity (OECD Series On Testing and Assessment, Number 15). The guidelines describe the systems currently in use by member organizations and countries including the European Union, United States, Canada, and Australia. They also provide overviews and comparisons of the systems in use, hazard classification methods, data requirements, mechanisms for classifying substances, and regulatory consequences of classification. An attachment lists substances classified in the European Union as toxic to reproduction.

In the United States, test guidelines for developmental toxicity testing are designed to provide general information concerning the effects of exposure of pregnant test animals on developing organisms. The guidelines were published by the EPA and meet the labelling requirements of the Federal Hazardous Substances Act (FHSA), the Federal Insecticide, Fungicide and Rodenticide Act (FIFRA) and the Toxic Substances Control Act (TSCA). Developmental neurotoxicity testing guidelines are also designed to address functional and behavioral effects of prenatal exposure, including teratogenicity and embryotoxicity information.

Reproduction and fertility testing is designed to incorporate gonadal function, the estrus cycle, mating behavior, conception, gestation, parturition, lactation and weaning, and growth and development of offspring. Proposed revisions to the developmental and reproductive toxicity testing guidelines should increase the sensitivity of the studies in identifying other effects such as cartilaginous development in fetuses, spermatogenesis, and sexual maturation.

In general, the regulatory systems are broadly similar, are based on classifications of chemicals according to their intrinsic properties, and involve the consideration of reproductive effects seen in humans and those demonstrated in suitable animal tests. Some differences among the systems include coverage of specific reproductive effects, number of classification categories for each type of reproductive effect, the possibility of classifying on the basis of limited evidence, and concentration cut-offs for classifying mixtures. Labelling requirements may also pose formidable obstacles to harmonization of classification.

9.7 MALE REPRODUCTIVE TOXICOLOGY

9.7.1 SPERMATOGENESIS

The testis is composed of a series of highly convoluted seminiferous tubules enclosed in the tunica albuginea, tunica vaginalis, and supporting connective tissue and vasculature. The tubules are composed of three cell types: spermatogenic (germ) cells, Sertoli cells, and the interstitial cells of Leydig. As the germ cells proliferate during spermatogenesis, they undergo meiotic division and migrate from the basement membrane of the tubule toward the lumen.

Development of germ cells arises with the appearance of spermatogonia, embryonic stem cells, near the basement membrane. As spermatogonia continue their progression toward the tubular lumen, they differentiate to primary spermatocytes, secondary spermatocytes, and spermatids. Spermiogenesis follows for an additional 5 wk in humans and involves the differentiation of spermatids into mature spermatozoa. Sperm cells undergo a metamorphosis from rounded cells that adhere onto the apical surfaces of seminiferous tubules into the characteristically shaped mature spermatozoa.

Spermiation is the final stage of differentiation that allows for the release of mature spermatozoa into the efferent ductules. Final transport of mature sperm involves traversing an extensive route through the reproductive ducts with eventual storage in the seminal vesicles.

Spermatozoa are continuously produced throughout the male reproductive life. The time for development, differentiation, and release of sperm varies among species. Table 9.3 outlines the features of male spermatogenesis according to species and also lists factors that influence spermatogenesis within species or strains.

The physical requirements for spermatogenesis, spermiogenesis, and spermiation are supported by the Sertoli cells, while endocrine maintenance is provided by the Leydig cells with the secretion of testosterone. In addition, endocrine control of these processes is regulated through the release of follicle stimulating hormone (FSH) and luteinizing hormone (LH) from the anterior pituitary. The releases of FSH and LH are in turn regulated by negative feedback inhibitory processes from the hypothalamic release of gonadotropin regulatory hormone (GnRH) as well as by direct inhibition of any further releases of anterior pituitary FSH and LH by their

TABLE 9.3
Differential Features of Male Spermatogenesis

Male Reproductive Characteristic	Mouse	Rat	Dog	NZ Rabbit	Human
Age at puberty (wk)[a]	4 to 6	6	32	20	520
Weight of testes (g)	?	3.5	13	6.4	34
Length of one cycle of seminiferous epithelium (days)	8.9	12.9	13.6	10.7	16
Features of collected sperm:					
Volume (ml)	NA	NA	3.0	0.5	1.5
Sperm concentration[b]	NA	NA	100	300	100
Total sperm per ejaculate[b]	NA	NA	300	160	150
Sperm output/g testes[b]	NA	NA	23	25	4.4
Percent sperm motility	NA	NA	80	80	50
Sperm capacitation (hr)	NA	NA	4	6	5
Natural mating	+	+	+	+	+
Sexual behavior	+	+	+	+	+

[a] Production of spermatozoa.
[b] 10^6.
Source: Adapted from Foote, R.H. and Carney, F.W., *Reprod. Toxicol.*, 14, 477, 2000.

own circulating blood levels. Figure 9.2 illustrates the endocrine feedback control mechanisms governing hormonal regulation of female gametogenesis.

9.7.2 REPRODUCTIVE TESTS

The meiotic process that governs sperm development is physiologically, morphologically, and genetically a vulnerable stage of the male reproductive cycle. Each stage of division is unique and may occur simultaneously within different parts of the seminiferous tubules. Thus, although a toxic insult has the capacity to affect any cell during a particular stage, it is also less likely that all cells in the testes will be affected equally. In addition, protection from xenobiotics and external influences is conferred by the formation of a blood–testis barrier via the Sertoli cells. Consequently, the toxicokinetic principles that govern passage of chemicals through phospholipid membranes also oversee penetration of testicular barriers.

In spite of these protective effects, genetic causes account for 10 to 15% of severe human male infertility including chromosomal aberrations and single gene mutations. In addition, the impact of cytotoxic chemotherapy on gametogenesis has demonstrated significant cytostatic and dose-specific responses. Therefore, in order to evaluate and understand the mechanisms related to the role of xenobiotics on male gametogenesis, several indicators of male reproductive toxicity are available to detect toxic insult. They include tests for sperm production, viability, morphology and experiments such as the dominant lethal assay intended to determine toxic effects

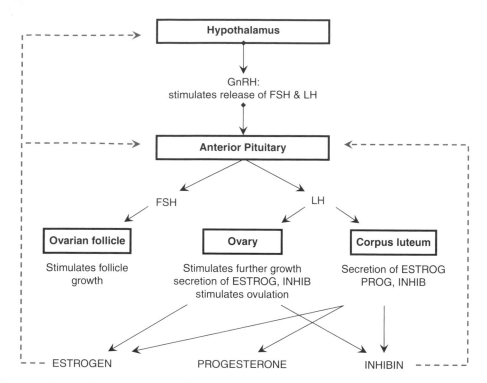

FIGURE 9.2 Endocrine feedback control mechanisms for gonadotropin releasing hormone (GnRH), follicle stimulating hormone (FSH), and luteinizing hormone (LH). Dashed lines indicate negative feedback inhibition. ESTROG = estrogen. INHIB = inhibin. PROG = progesterone.

of chemicals on mating, sexual behavior, and fertility and pregnancy outcomes. Table 9.4 lists a variety of tests used for assessing male reproductive toxicity (some of the parameters incorporated into male reproductive toxicity tests are already listed in Table 9.3).

Several effects may be determined, depending on the times and sequences of responses. For instance, an acute reduction in the fertility index after treatment suggests an adverse effect on mature spermatozoa, while the discovery of delayed effects indicates interference with spermatogenesis. Morphological examination of the components of the testes and tubules are included to confirm the reproductive data. Acute and extended mating studies of rodents and rabbits require one or several turnover cycles of the seminiferous epithelium, resulting in several weeks to months of treatment, in order to achieve a steady-state between agent and target site.

Combinations of these assays are capable of screening both strong and weak toxicants based on dose–response parameters. Anti-fertility effects, influence on implantation, pre- and post-spermatogenic effects, and fetal lethality are assessed with appropriately planned and executed timing of protocols. Consequently, based on evaluation of the results, mechanistic information is achievable. Conclusions about toxic effects are derived from direct action on the meiotic process

TABLE 9.4
Recommended Male Reproductive Toxicity Tests

Test Parameter	Features
Sperm production	Measurements of ejaculatory volume, sperm concentration, total sperm count, motility, morphology, capacitation
Serial mating	Males treated with single or repeated doses of agent and mated continuously with untreated females
Extended mating studies	Subchronic administration of test agent over several spermatogenic cycles prior to serial mating with untreated female animals
Other specific parameters:	
Fertility Index	Percentage of matings resulting in successful pregnancy during the time period
Dominant lethal assay	Single, repeated, acute or chronic exposure of males throughout a complete spermatogenic cycle, then serially mated with untreated females; objective: monitoring for severe chromosomal damage or lethal germ cell mutations

to the influence on the neuroendocrine system, including hypothalamic–hypophyseal feedback, effects on autonomic regulation of erection and ejaculation, and mating behavior.

While these protocols provide valuable information for the determination of potential reproductive toxicity, the methods for conducting the assessments have not been well developed in all laboratories and are continually evolving. The lack of standardization of methods does not allow for smooth transition of data among laboratories. To address the differences in methods, however, several recent working groups have convened to discuss optimization of methods for assessing sperm parameters in animal species. The Suggested Readings and Review Articles section at the end of this chapter provides more information and lists summaries and outlines of individual protocols.

9.8 FEMALE REPRODUCTIVE TOXICOLOGY

9.8.1 OOGENESIS

The ovaries are the paired female gonads that contain primordial follicles in which oogenesis occurs. Oogenesis is defined as the development and formation of mature secondary oocytes. The follicles arise from primordial germ cells and represent the corresponding embryonic stem cells in the female reproductive cycle. The meiotic process is more complex in female mammals, principally because of the interruptions in cycles.

Meiosis I begins during fetal life and stops at birth with the development of oogonia. A genetically determined number of oogonia develop and remain dormant until puberty when they are recruited to continue with the monthly meiotic divisions. In fact, oogonia represent the entire allotment of germ cells available for meiotic reproduction during female adult life and are consumed at an approximate rate of 100 to 200 primordial follicles per menstrual cycle. At puberty, primary oocytes complete meiosis I, which produces a secondary oocyte and a first polar body that may divide again. At this stage, the secondary oocyte is in the haploid (N) state and is surrounded by the zona pelucida and granulosa cells within the developing secondary follicle.

The secondary oocyte then begins meiosis II, is arrested at metaphase of meiosis II, and is ovulated with a pronucleus (haploid number of chromosomes) and captured in the fallopian tube. Ovulation occurs in response to a luteinizing hormone (LH) "surge." The process results in follicle rupture, discharge, and capture of the ovum by the fallopian tube. The corpus luteum forms from the collapse of the ruptured follicle and secretes progesterone.

After fertilization, the secondary oocyte splits into a mature ovum along with a second polar body. The pronuclei of the sperm cell and ovum fuse and form a diploid (2N) zygote. The next immediate stages are characterized by differentiation of the zygote to the phases of the morula and blastocyst, the latter of which proceeds to implantation within 7 to 10 days.

As with the male reproductive cycle, the menstrual cycle and development of the secondary oocyte are regulated by FSH and LH from the anterior pituitary, while developing follicles and corpora lutea secrete estrogens and progesterone, respectively. The hormones regulate feedback mechanisms that further control ovarian functions (see Figure 9.2). Similarly as with male animals, female species vary considerably in time and sequence of oogenesis. Table 9.5 outlines a number of parameters for female laboratory species, in comparison to humans.

TABLE 9.5
Differential Features of Female Oogenesis

Female Reproductive Characteristic	Mouse	Rat	Dog	NZ Rabbit	Human
Age at puberty (wk)[a]	4 to 6	7 to 10	7 to 9	24	520
Age at sexual maturity (wk)	6 to 8	9 to 14	10 to 14	32	624
Adult weight (g)	25 to 80	250 to 300	10 kg	5000	55 kg
Length of menstrual cycle (days)	4 to 6	4 to 5	Variable	Variable	28
Length of estrus (hr)	10 to 14	10 to 16	>96	—	—
Number of oocytes at ovulation	8 to 12	10 to 14	10	9 to 14	1 to 2
Time of ovulation (hr)	2 to 3	8 to 10	63	31	~24
Length of gestation (days)	19 to 21	22	63	31	260
Litter size (number)	6 to 12	8 to 12	4 to 8	7 to 10	1 to 2
Reproductive life (yr)	1	1	14	3 to 5	30 to 35

[a] Production of spermatozoa.

Source: Adapted from Foote, R.H. and Carney, F.W., Reprod. Toxicol., 14, 477, 2000.

TABLE 9.6
Assessment Parameters of Female Reproductive Toxicity Tests

Parameter	Features	Surgical Manipulation
Endocrine status	Hypophyseal–hypothalamic– pituitary integrity in association with normal levels of gonadotropins	No
Blastocyst recovery	Indication of fertilization and preimplantation development; rapid mitotic division	No
Determination of fertilization	Measure sperm attached to surrounding oocyte or young embryo in rabbit	Yes
Determination of sperm in mucin coat	Number of sperm trapped in mucin coat is enumerated as indication of sperm transport	Yes
Cleavage rate	Monitored *in vitro* for several days post-fertilization; monitoring of mitotic divisions	Yes
Implantation	Hystological examination	Yes
Fetal development	Development of germ layers: endoderm, ectoderm, mesoderm	Yes
Reproductive tract morphology	Macroscopic and histological examination	Yes
Rate of embryo transport to oviduct	Indicator for embryo maturation and endometrial development	Yes

Source: Adapted from Foote, R.H. and Carney, F.W., *Reprod. Toxicol.*, 14, 477, 2000.

9.8.2 REPRODUCTIVE TESTS

As with the male reproductive cycle, the meiotic process that governs oogenesis is physiologically, morphologically, and genetically more vulnerable to chemical insult than spermatogenesis. The meiotic process is more uniform; the stages of differentiation are sequential and usually do not occur with more than one ovum simultaneously within the ovary. Consequently, it is highly likely that a toxic insult has the capacity to profoundly affect development of mature ova at any time during the menstrual cycle.

In order to assess the mechanisms related to the role of xenobiotics on female gametogenesis, several indicators of female reproduction are outlined in Table 9.6. These features represent important parameters in the evaluation of normal reproductive status and can then be monitored after administration of a toxic substance.

Table 9.7 summarizes a variety of recommended female reproductive toxicity test combinations that are capable of screening both strong and weak toxic substances based on dose–response parameters. Anti-fertility effects, influence on female gametogenesis, implantation, effects on pre- and post-organogenesis, and fetal abnormalities are assessed with appropriately planned and executed timing of protocols. Conclusions on toxic effects are derived from analysis of the steps involved in female reproduction.

TABLE 9.7
Recommended Female Reproductive Toxicity Tests

Parameter	Features	Indicators
Single generation study (fertility and teratogenicity studies)	Agent administered to female animals for ~15 days prior to starting breeding program; treatment continues throughout pregnancy	Determination of fertility index[a], mating index[b], gestation index[c]; histopathologic and microscopic examinations
Multiple generation studies	Treatment as per single generation study until newborns reach reproductive status; studies monitored over 2 to 3 generations	Behavioral, physiological, and biochemical anomalies displayed in offspring as consequence of maternal exposure
Post-parturition studies	Treatment as per single generation studies	Behavioral, rearing, lactation, sensory, motor, and cognitive neurological function

[a] Ratio of number of females conceiving divided by number of females exposed to proven, fertile males is a valuable indicator of overall reproductive capacity.
[b] Number of estrus cycles required of each female to produce pregnancy and number of copulations.
[c] Number of pregnancies resulting in births of live litters.

TABLE 9.8
Segment Testing

Segment	Protocols
I	Separate treatment of males or females with test substance, but not both or together
II	Normal fertilization: untreated females mated with untreated males; administration of agent post-fertilization
III	Assessment for teratogenic and embryotoxic effects during prenatal development; examination of post-natal development including effects on parturition and fetal survival, lactation, weaning, and maturity

Finally, Table 9.8 outlines a description of *segment testing* and assessment of reproductive toxicity. This sequential conduct of studies enables a systematic evaluation of toxicology testing data while minimizing unnecessary duplication of laboratory and animal resources.

SUGGESTED READINGS

Anonymous, International Conference on Harmonisation: guidelines on detection of toxicity to reproduction for medicinal products, *Fed. Reg.*, 59, 48746, 1994.
Flick, B. and Klug, S., Whole embryo culture: an important tool in developmental toxicology today, *Curr. Pharm. Des.*, 12, 1467, 2006.

Hurtt, M.E., Cappon, G.D., and Browning A., Proposal for a tiered approach to developmental toxicity testing for veterinary pharmaceutical products for food-producing animals. *Food Chem. Toxicol.*, 41, 611, 2003.

Organization of Teratogen Information Services, http://www.otispregnancy.org/, last accessed November 20, 2006.

U.S. Environmental Protection Agency, Health Effects Test Guidelines, OPPTS 870.3700, Prenatal Developmental Toxicity Study, EPA 712, 207, 1998.

REVIEW ARTICLES

Bignami, G., Economical test methods for developmental neurobehavioral toxicity, *Environ. Health Perspect.*, 104, 285, 1996.

Brachet, A. Recherches sur le determinism hereditaire de l'ouef des manniferes: development in vitro de jeunes vesicules blastodermique du lapin, *Arch. Biol.*, 28, 447, 1913.

Buschmann, J., Critical aspects in reproductive and developmental toxicity testing of environmental chemicals, *Reprod. Toxicol.*, 22, 157, 2006.

Cohen, S.M., Robinson, D., and MacDonald, J., Alternative models for carcinogenicity testing, *Toxicol. Sci.*, 64, 14, 2001.

Detailed Review Document on Classification Systems for Reproductive Toxicity in OECD Member Countries, OECD Series on Testing and Assessment, Number 15, 1999.

Foote, R.H. and Carney, E.W., The rabbit as a model for reproductive and developmental toxicity studies, *Reprod. Toxicol.*, 14, 477, 2000.

Hurtt, M.E., Cappon, G.D., and Browning A., Proposal for a tiered approach to developmental toxicity testing for veterinary pharmaceutical products for food-producing animals, *Food Chem. Toxicol.*, 41, 611, 2003.

Jelinek, R., The contribution of new findings and ideas to the old principles of teratology, *Reprod. Toxicol.*, 20, 295, 2005.

Riecke, K. and Stahlmann, R., Test systems to identify reproductive toxicants. *Andrologia*, 32, 209, 2000.

Seed, J., et al., Methods for assessing sperm motility, morphology, and counts in the rat, rabbit, and dog: a consensus report, *Reprod. Toxicol.*, 10, 237, 1996.

Storgaard, L., Bonde, J.P., and Olsen, J., Male reproductive disorders in humans and prenatal indicators of estrogen exposure: a review of published epidemiological studies, *Reprod. Toxicol.*, 21, 4, 2006.

Tam, P.P., Post-implantation mouse development: whole embryo culture and micro-manipulation, *Int. J. Dev. Biol.*, 42, 895, 1998.

Walker, R., The significance of excursions above the ADI: duration in relation to pivotal studies, *Regul. Toxicol. Pharmacol.*, 30, S114, 1999.

Webster, W.S., Brown-Woodman, P.D., and Ritchie, H.E., A review of the contribution of whole embryo culture to the determination of hazard and risk in teratogenicity testing, *Int. J. Dev. Biol.*, 41, 329, 1997.

10 Carcinogenicity and Mutagenicity Testing *In Vivo*

10.1 INTRODUCTION AND DEFINITIONS

Cell proliferation occurs in most tissues throughout a lifespan and is influenced by a variety of circumstances. During this normal physiologic state, the delicate balance between cell proliferation and apoptosis (programmed cell death) is perpetuated to ensure the integrity and proper function of organs and tissues. Mutations in DNA that lead to cancer development interfere with this orderly process by disrupting its regulation. Consequently, the initiation of a cancer is due to an abnormal and uncontrolled progression of cell proliferation characterized by unregulated cell division and metastasis (spreading) of foci of cells to distant tissues.

Carcinogenesis represents the unwarranted appearance or increased incidence of abnormal cell proliferation occurring in an age-matched low-risk group whose propensity for development of cancer is statistically lower. A carcinogen is any chemical or viral agent that increases the frequency or distribution of new tumors, results in their appearance within a low-risk or otherwise early age group, or results in the introduction of new pathological growths otherwise absent in experimental controls.

Most chemical carcinogens require metabolic activation before demonstrating carcinogenic potential. As with most toxic phenomena, a minimum dosage is necessary to elicit a carcinogenic event. *Epigenetic* or *non-genotoxic* carcinogens enhance the growth of tumors by mechanisms other than through alteration of DNA. These chemicals influence absorption, biotransformation, or reduce elimination of the initiating agent. *Co-carcinogens* redirect hormonal action on cell proliferation and inhibit intercellular communication, thus allowing for greater proliferative capacity. Alternatively they induce immunosuppression of protective immunological pathways.

The carcinogenic process is on occasion a response to a mutation occurring within the genetic materials of normal cells, resulting in uncontrolled cell division and transformation to the immortal phenotype. The unrestrained and often rapid proliferation of cells is further characterized as benign or malignant tumors. *Benign tumors* do not metastasize, are usually confined to local target organ areas, are clincally susceptible to therapeutic intervention, and carry more favorable prognoses. Alternatively, *malignant tumors* metastasize to distant organ locations, are not necessarily amenable to therapeutic intervention, and possess less favorable prognoses.

Mutagenesis refers to the ability of a virus or chemical agent to induce changes in the genetic sequences of mammalian or bacterial cells, thus altering the phenotypic expression of cell characteristics. *Genotoxicity* refers to the ability of an agent to induce heritable changes in genes that exercise homeostatic control in somatic cells while increasing the risk of influencing benign or malignant transformation. Genotoxic substances induce genotoxicity by either binding directly to DNA or by indirectly altering the DNA sequence, causing irreversible damage. It is also important to note, however, that genotoxic substances are not necessarily carcinogenic.

In addition, interaction of a chemical, physical, or viral agent with nucleic acids results in the disruption of the transfer of genetic information and the development of genotypic and phenotypic consequences. Lastly, *mitogenesis* is the induction of cell division (mitosis) within eukaryotic or prokaryotic cells by stimulating transit through the cell cycle. Prolonged and continuous exposure to growth factors is required to commit cells to the cell cycle.

10.2 MULTISTAGE CARCINOGENESIS

The process of carcinogenesis is divided into three experimentally defined stages: *tumor initiation*, *tumor promotion*, and *tumor progression*. This multistage development requires the malignant conversion of benign hyperplastic cells to a malignant state, involving invasion and metastasis as manifestations of further genetic and epigenetic changes.

10.2.1 TUMOR INITIATION

The early concept of tumor initiation indicated that the initial changes in chemical carcinogenesis involved irreversible genetic alterations. Recent data from molecular studies of pre-neoplastic human lung and colon tissues, however, implicate epigenetic changes as an early event in carcinogenesis. DNA methylation of promoter regions of genes can transcriptionally silence tumor suppressor genes. Thus, carcinogen–DNA adduct formation is central to the theory of chemical carcinogenesis and may be a necessary but not sufficient prerequisite for tumor initiation.

DNA adduct formation that causes either the activation of a proto-oncogene or the inactivation of a tumor suppressor gene is categorized as a tumor initiating event. One important characteristic of this stage is its irreversibility: the genotype or phenotype of the initiated cell is conferred during the process. Chemicals capable of initiating cells are referred to as initiating agents. Furthermore, initiation without the subsequent promotion and progression rarely yields malignant transformation.

10.2.2 TUMOR PROMOTION

In multistage carcinogenesis, tumor promotion comprises the selective clonal expansion of initiated cells through a mechanism of gene activation. Because the accumulation rate of mutations is proportional to the rate of cell division, clonal expansion of initiated cells produces a larger population of cells that are at risk of further genetic changes and malignant conversion. Tumor promoters are generally non-mutagenic, are not solely carcinogenic, and often are able to mediate their biologic effects without metabolic activation. In addition, they do not directly interact with DNA.

Interestingly, tumor promotion is reversible, although the continued presence of the promoting agent maintains the state of the promoted cell population (pre-neoplastic lesion). Thus the stage appears to have a long duration, especially in humans, and is a preferred target for experimental manipulation. Examples of typical tumor promoters include tetradecanoyl phorbol acetate (TPA), phenobarbital, and 2,3,7,8-tetrachlorodibenzo dioxin (TCDD).

During tumor promotion, *malignant conversion* may occur in which a pre-neoplastic cell is transformed into the malignant phenotype. The process requires further genetic changes. Frequent repeated administration of the tumor promoter is more important than the total dose. In addition, if the tumor promoter is discontinued before malignant conversion has occurred, pre-malignant or benign lesions may regress. Tumor promotion contributes to the process of carcinogenesis by the expansion of a population of initiated cells that are then at risk for malignant conversion. Conversion of a fraction of these cells to malignancy is accelerated in proportion to the rate of cell division and the quantity of dividing cells in the benign tumor or pre-neoplastic lesion. In part, these genetic changes result from infidelity of DNA synthesis.

10.2.3 TUMOR PROGRESSION

Tumor progression comprises the expression of the malignant phenotype and the tendency of malignant cells to acquire more aggressive characteristics over time. Also, metastasis may involve the ability of tumor cells to secrete proteases that allow invasion beyond the immediate primary tumor location. A prominent characteristic of the malignant phenotype is the propensity for genomic instability and uncontrolled growth. Further genetic and epigenetic changes occur, including the activation of proto-oncogenes and the functional loss of tumor suppressor genes.

Loss of function of tumor suppressor genes usually occurs in a bimodal fashion, and most frequently involves point mutations in one allele and loss of the second allele by a deletion, recombinational event, or chromosomal non-disjunction. These phenomena confer growth advantage to the cells along with the capacity for regional invasion, and ultimately distant metastatic spread. Despite evidence for an apparent scheduling of certain mutational events, the accumulation of these mutations and not the order or the stage of tumorigenesis in which they occur appears to be the determining factor.

Figure 10.1 illustrates the events that occur during the three stages of chemical carcinogenesis and the requirements necessary for the completion of the multistage process.

10.3 CARCINOGENIC AND GENOTOXIC AGENTS

Both collective and individual exposures often define environmental toxicology. Epidemiological research supported by experimental investigations led to the determination of profiles of various risk factors. These profiles are perpetually incomplete because humans incessantly modify the environment, thus increasing the risks

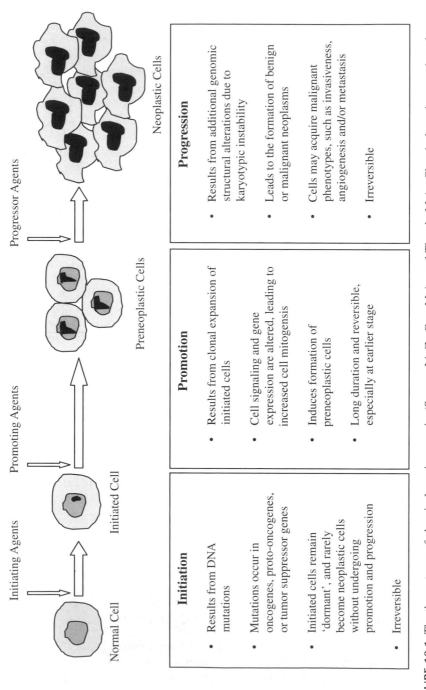

FIGURE 10.1 The three stages of chemical carcinogenesis. (Source: Li, Z., Cao, M.A., and Thrush, M.A., Chemical carcinogenesis and mutagenesis, in *Clinical Toxicology: Principles and Mechanisms*, Barile, F.A., Ed., CRC Press, Boca Raton, FL, 2004, Ch. 29. With permission.)

of interactions with agents against which there may be no protective mechanisms. The multiplicity of chemical and physical carcinogenic agents does not render them easy to study since the development of carcinoma is in all probability not due to contact with a single carcinogenic factor but rather to the sum of many possible causative agents.

Of the various environmental hazardous compounds, cigarette smoke enjoys the highest causal relationship with cancer risk in humans. Tobacco smoking plays a major role in the etiology of lung, oral cavity, and esophageal cancers and a variety of other chronic degenerative diseases. Although cigarette smoke is a mixture of about 4000 chemicals including more than 60 known human carcinogens, 4-methylnitrosamino-1-(3-pyridyl)-1-butanone (nicotine-derived nitrosamino ketone, NNK) is the most carcinogenic tobacco-specific nitrosamine. NNK induces lung tumors in mice, rats, and hamsters. The International Agency for Research on Cancer has designated NNK and NNN (N-nitrosonornicotine) as known human carcinogens. NNK is metabolically activated by CYP P-450 enzymes in the lungs and generates O6-methylguanine in DNA. The reaction generates G:C to A:T mutations with the subsequent activation of the K-ras proto-oncogene and development of tumor initiation.

Radiation also plays a significant role as one of the most causative physical factors inducing human cancers. Radiation promotes double-strand breaks (DSBs) in DNA that lead to chromosome aberrations and cell death and also generates a variety of oxidative DNA damage. Because of genotoxicity, radiation at high doses evidently results in the appearances of various tumors in humans. Even at low doses, residential exposure to radioactive radon and its decay products, for instance, may account for about 10 to 20% of all lung cancer deaths worldwide.

10.4 CARCINOGENICITY TESTING *IN VIVO*

10.4.1 PRINCIPLES OF *IN VIVO* CARCINOGENICITY TESTING

The toxicology community generally agrees that carcinogenicity testing and its application to human carcinogenic risk assessment need improvement. In addition, more information is necessary about the incorporation of mechanisms and modes of action into the risk assessment process. Advances in molecular biology have identified a growing number of proto-oncogenes and tumor suppressor genes that are highly conserved across species and are associated with an extensive variety of mammalian cancers.

In vivo transgenic rodent models that incorporate these mechanisms are used to identify pathways involved in tumor formation. Transgenic methods are considered extensions of genetic manipulation by selective breeding — a technique that has long been employed in science and agriculture. With carcinogenicity testing, the use of two rodent species is especially important for identifying trans-species carcinogens. The capacity of a substance to induce neoplastic events among species suggests that the carcinogenic mechanisms are conserved and therefore may have significance for humans. Based on available information, there is sufficient experience with some *in vivo* transgenic rodent carcinogenicity models to support their application as complementary second species studies in conjunction with a single 2-year rodent carcinogenicity study. Examples of assessment factors in the consideration of the adequacy of the test model and test compound include:

1. Adequacy of doses included in the study and suitability of the route of administration
2. Bioavailability of the test compound in rodent models
3. Toxicokinetic profile of the compound in rodent models and humans
4. Biotransformation profiles of the compound in rodent models and humans
5. Appropriate application of International Conference on Harmonization (ICH) guidelines

Risk-benefit considerations include:

1. Comparison of systemic exposure at estimated maximum daily recommended dose to systemic exposure in rodent test models.
2. If systemic exposure information is lacking, the estimated exposure ratio should be based on surface area (mg/m^2) estimation of dose in the test species and the maximum daily recommended clinical dose.
3. Severity of malignancy.
4. Size and composition of patient population and frequency and extent of exposure.

Even the smallest *in vivo* carcinogenicity studies, when performed properly, are expensive and time consuming. Consequently, only a relatively small fraction of new and existing chemicals has been tested accordingly. Therefore, the development and validation of alternative approaches to offset the resources necessary for *in vivo* identification and classification of human carcinogens and mutagens are important parts of carcinogenicity testing. Current methods used for this evaluation, however, such as the Ames test and long-term *in vivo* studies, suffer from several shortcomings, including:

1. The need to extrapolate the results of the Ames test from prokaryotes to humans
2. The need to extrapolate *in vivo* testing results from rodents to humans
3. The prolonged time required for the conduction of *in vivo* tests

The current methods also suffer from limited validation comparing the agreement of various methods with each other and applying laboratory results to human populations.

In 1996, the National Institute of Environmental Health Sciences (NIEHS, Research Triangle Park, North Carolina) initiated the predictive toxicology evaluation (PTE) project to develop and validate improved methods to predict the carcinogenicities of compounds. The project conducts collaborative experiments that subject the performance of predictive toxicology (PT) methods to objective evaluation and incorporate ongoing testing conducted by the National Toxicology Program

(NTP) to estimate the true errors of models applied to make prospective predictions on previously untested chemical substances.

The study first describes the identification of a group of standardized NTP chemical bioassays, then announces and advertises the evaluation experiment. The information about the bioassay is distributed to encourage researchers to publish their predictions in peer-reviewed journals, using unsolicited approaches and methods, after which the collection of data is published to compare and contrast PT approaches and models to applied situations. Several chemicals carcinogenesis bioassays have been presented since the first supplement publication of the PTE, along with a table of prediction sets, challenges associated with estimating the true error of models, and a summary.

10.4.2 *In Vivo* Genotoxicity and Cytogenetic Tests

Several validated *in vivo* models are currently accepted for assessment of genotoxicity. These include the bone marrow or peripheral blood cytogenetic assay. When a compound demonstrates negative results *in vitro*, performance of a single *in vivo* cytogenetics assay is usually sufficient. A compound that induces a biologically relevant positive result in one or more *in vitro* tests requires further testing with an *in vivo* model in addition to the cytogenetic assay. The test usually incorporates a tissue other than the bone marrow or peripheral blood that can provide further clarification of effects. Compounds that screen positively in any of the *in vitro* tests, particularly the bone marrow cytogenetic assay, require confirmation with an *in vivo* method accordingly:

1. Determining any significant change in the proportion of immature erythrocytes among total erythrocytes in the bone marrow at the doses and sampling times used in the micronucleus test or measuring a significant reduction in mitotic index for the chromosomal aberration assay
2. Providing evidence of bioavailability of substance related material by measuring blood or plasma levels
3. Directly measuring substance-related material in bone marrow
4. Performing autoradiographic assessment of tissue exposure

To date, no validated, widely used *in vivo* system measures gene mutation. *In vivo* gene mutation assays using endogenous genes or transgenes in several tissues of rats and mice are at various stages of development and until such tests for mutation are accepted, results from other *in vivo* tests for genotoxicity in tissues other than the bone marrow can provide valuable additional data. Disagreements between *in vivo* and *in vitro* test results performed in parallel must be analyzed separately.

10.4.3 Conclusions

The assessment of the genotoxic and carcinogenic potential of a compound should examine all of the relevant findings and acknowledge the intrinsic values and limitations of both *in vitro* and *in vivo* tests.

In vivo tests have an important role in genotoxicity and carcinogenicity testing strategies. The significance of *in vivo* results in genotoxicity testing strategies is proportional to the demonstration of adequate exposure of the target tissue to the test compound. This is especially true for negative *in vivo* test results and when *in vitro* tests show convincing evidence of genotoxicity. In addition, a dose-limiting toxic response can occur in a tissue other than the genotoxic target organ. In such cases, toxicokinetic data is used to provide evidence of bioavailability. If adequate exposure is not achieved, for example with compounds demonstrating poor target tissue availability, extensive protein binding and other conventional *in vivo* geno-toxicity tests may be of limited value. Details of available *in vitro* carcinogenicity and mutagenicity assays are discussed in Chapter 15.

10.5 BIOASSAY PROTOCOLS FOR *IN VIVO* CARCINOGENICITY TESTING

10.5.1 STUDY DURATION

Based on the mechanisms of carcinogenesis described above and the unpredictable nature of dose–response relationships in the induction of tumors, *in vivo* carcinoge-nicity studies are necessarily performed for 18 to 24 mo. European Union guidelines specify a minimum of 18 mo for mice and 24 mo for rats; United States and OECD regulations require studies for at least 18 to 24 mo in mice and 24 to 30 mo in rats or for the lifespan of a species in the event of a high survival rate.

These chronic assays ensure that the time interval of exposure to a potential carcinogen is sufficient to induce a carcinogenic response within the major part of a rodent's lifespan. The studies generally combine qualitative data with quantitative results for human risk assessments, the purpose of which is to demonstrate that a chemical lacks the ability to induce or promote cancer formation.

10.5.2 INDICATORS FOR DETECTION OF CARCINOGENS

In contrast to acute or chronic toxicity tests, animal carcinogenicity assays rely on what appears to be a single endpoint: the development of neoplasia. Consequently, more attention is focused on the morphological examinations of organs and tissues in an effort to evaluate the carcinogenic response. The criteria include (1) determin-ing the onset of tumors, (2) evaluating for appearance and number of different types of tumors, and (3) determining the number of tumor-bearing animals.

Various physiological and toxicological indicators are also monitored throughout the length of a study when economically feasible, especially monitoring of growth and development, food and water consumption, mortality rate, morbidity, and inter-mittent hematological and biochemical evaluation of blood and urine samples. Rep-resentative target and non-target organs are extirpated from animals periodically and at study termination and subjected to morphological and histological studies. Other specimens are also processed for clinical evaluation.

10.5.3 SELECTION OF ANIMAL SPECIES AND NUMBERS

Similar factors that affect the selection of animals in chronic toxicity studies (Chapter 7) also influence the selection of animal models for carcinogenic projects. In general, regulatory guidelines require 50 animals of each sex per treatment group, preferably using both species of rodents. The mouse and the rat are the species of choice, principally because of the cumulative favorable factors that they afford to chronic studies: availability, relatively fewer resources needed during the length of the study, smaller size of animals and ease of handling, routine housing requirements, shorter lifespans that more closely mimic chronic exposure, and an abundance of literature available on their carcinogens.

10.5.4 EXPOSURE LEVELS

The *maximum tolerated dose* (MTD) of chemical exposure without demonstration of adverse non-carcinogenic effects is selected as the upper limit dose. Two proportionately linear or logarithmic lower doses plus controls complete the treatment groups. In a 12-week preliminary investigation, the MTD should not cause losses of body weight greater than 10% in comparison to matched control groups. In addition, the MTD should not cause significant mortality, clinical signs of toxicity, or pathologic lesions other than those related to carcinogenicity among the treatment groups. Animals in the control groups are also monitored for normal frequency of tumor incidence, especially toward the latter stages of a chronic study.

Although the MTD is associated with several controversial features that render the results difficult to interpret, it is still a feasible general guideline for establishing dosage levels. One of the major problems associated with the assignment of the MTD as the upper limit dose is that by attempting to correlate the interaction between a toxicologic response and a carcinogenic observation in relation to assigned dose, the individual mechanisms may not have enough commonality to expect a carcinogenic phenomenon even when the dose is toxicologically tolerated.

Alternatively, in the absence of a noticeable toxic effect such as with a weak toxicant (e.g., food additive), the MTD selected may be exceedingly high so that the toxicological mechanism associated with the chemical at the high dose is responsible for the induction of a carcinogenic event. The situation is analogous to the induction of a carcinogenic effect by a mitogenic chemical.

Another important criterion in the selection of doses is to estimate the lowest dose at which carcinogenicity is not observed the no-observable-adverse-effect level (NOAEL). The objective of targeted exposure levels, therefore, is to estimate the incidence of tumor induction within a range of doses for a test chemical without producing significant non-carcinogenic (toxic) events.

The chemical is then administered through an appropriate route, usually either in drinking water or formulated as a component of diet over the duration of the study. Administration via the inhalation, gastric gavage, dermal, or parenteral routes is reserved for specialty studies.

10.5.5 Sources of Variability

Because of the extended duration and complexity of carcinogenicity studies, several factors may influence tumor incidence in experiments and contribute to inconsistent or unreliable interpretation of the data factors that may not be of concern in chronic toxicity studies. The condition of the animal care quarters, genetic differences among strains, amounts of food consumption and weight gain, survival ages of the animals, identification of gross lesions, pathology sampling procedures, preparation of the specimens for histological and morphological analysis are all factors that contribute to variability of results during a project. Evaluation of data depends on objective and subjective observations and analyses and statistical applications throughout the duration of a study.

Another complicating factor necessarily involves the significant background incidence of spontaneous tumor formations in rodents. Rodents demonstrate a marked predisposition toward the production of specific tumor types at particular sites, especially dermal, pulmonary, and hepatic neoplasms. This suggests that a rodent possesses an initiated genotype that biases the animal strain toward the formation of pre-neoplastic lesions.

Interestingly, new questions have surfaced about a particular feature of the design, conduct, and interpretation of the data from original carcinogenicity studies performed with animals fed *ad libitum* diets (food is available 24 hr/day without restriction throughout the course of a study). In the past 20 years, evidence indicates that animals on restricted calorie diets are leaner, have higher basal metabolic rates, increased survival times, decreased tumor incidence, and fewer pathological complications with age. The influence of calorie restriction on subchronic and chronic toxicology projects is discussed in detail in Chapter 7. These factors have similar consequences in the conduct of carcinogenicity programs.

10.5.6 Carcinogenicity Bioassays

Chapter 15 outlines a series of *in vitro* tests as screening methodologies to aid in the preliminary planning for chronic *in vivo* carcinogenicity studies. In general, three systems are currently in practice that allow for progression to long-term animal projects. These include (1) short-term test batteries, (2) tier testing, and (3) decision point analysis.

10.5.6.1 Short-Term Test Batteries

This popular approach includes the incorporation of a battery of short-term tests with different carcinogenic indicators. The use of assays with various mechanistic targets is the basis for the ability of the short-term tests to detect a greater range of compounds with carcinogenic potentials. Although the principle of the system appears sound, concern about the technical aspects of the various tests and applications of the results to human risk assessment requires cautious interpretation of data.

TABLE 10.1
Decision Point Analysis for Detection of Chemicals with Carcinogenic Potentials

Level	Procedure	Description
A	Structure–activity relationship	Analysis of structural features of a chemical and molecular similarities that suggest similar mechanisms of toxicity of chemicals in structurally related classes
B	Short term *in vitro* assays	Bacterial mutagenesis; mammalian mutagenesis; cell transformation; mammalian DNA repair assays
	Decision point 1: evaluation of results from tests at levels A and B	
C	Test for promoters	*In vitro* assays; *in vivo* assays
	Decision point 2: evaluation of results from tests at levels A, B, *and* C	
D	Limited *in vivo* bioassays	Rodent liver altered foci induction; mouse skin neoplasm induction; mouse pulmonary neoplasm induction; female Sprague-Dawley rat breast cancer induction
Decision point 3: evaluation of results from tests at levels A, B, C, *and* appropriate test from D		
E	Chronic bioassay	Chronic carcinogenic studies in rodents
Decision point 4: Final evaluation of results from tests at all levels for risk assessment analysis		

Source: Adapted from Williams, G.M. and Weisburger, J.H., Mutat. Res., 205, 79, 1988.

10.5.6.2 Tier Testing

The principle of tier testing relies on a systematic sequence of ranked levels used in a testing strategy: a positive screen at the first level qualifies for confirmation at subsequent levels. The Ames test has frequently been employed as a screening test system in the first tier, followed by performance of genotoxicity tests in the second and third tiers.

10.5.6.3 Decision Point Analysis

Table 10.1 offers a scheme for the decision point analysis for carcinogen testing. The approach is based on systematically delineating the mechanism of toxicity of a chemical. The sequential levels commence with structure–activity analysis of the chemical as a preliminary screen, followed by a battery of *in vitro* short-term tests. These tests are then evaluated at decision point 1. If results are negative or inconclusive, tests for carcinogenic promoters are continued, after which decision point 2 is determined. Depending on the conclusions obtained, the process continues through decision points 3 and 4. If results are definitive about the carcinogenic potential of a compound at any level, no further testing is needed. Interpretation of the results from combinations of the bioassays may involve testing of the chemical in chronic studies. Nevertheless, conclusions concerning risk assessment about carcinogenic potential are accomplished with a minimal number of animals and reduced number of protocols.

10.5.6.4   Rodent Liver Altered Foci Induction

The altered foci induction test system measures the regenerative capacity of rat liver following partial hepatectomy in the presence of initiator or promoter chemicals. The indicator relies on the formation of foci from rapidly replicating, phenotypically altered cells. The foci are suggestive of pre-neoplastic lesions that produce hyperplastic nodules or transform to malignant growths. In addition, the higher levels of phase I enzyme activity in the altered hepatocyte foci enable the biotransformation of procarcinogens. Histochemical, immunochemical, and histological assays are employed for the detection of chemical and enzyme biomarkers and pathology, respectively, that are associated with tumor induction. The technique is also modified to detect tumor promoters. Thus the protocol takes advantage of a marked reduction in the time period and an increase in the sensitivity required for the demonstration of carcinogenic initiation and promotion.

10.6   MUTAGENICITY TESTING *IN VIVO*

Mutations are hereditary genotypic changes that occasionally express phenotypic alterations. Thus a defect in the nucleotide sequence results in misreading of the genetic code. In the absence of adequate DNA repair mechanisms, the consequence of such a mutation may result in carcinogenic sequelae. The spectrum of genetic damage is dependent on those factors delineated above in the conduction of carcinogenic studies, including level and duration of exposure, extent of genotypic alteration, and response of the organism to the initiation event.

The bases of tests for mutagenic potential are discussed in detail in Chapter 15. Briefly, the bacterial tests incorporate the ability to detect reverse mutations from the nutrient-dependent strain to a "wild-type" capable of sustaining itself on minimal medium or chemically induced forward mutations resulting in measurable phenotypes. Forward mutations occur at several loci within one gene, represent a larger target for mutagenic chemicals, and consequently are relatively easier to detect. Reverse mutation assays rely on the abilities of organisms to induce mutations at a specific locus, thus providing a smaller selective target site for a chemical.

SUGGESTED READINGS

Ames, B.N., Identifying environmental chemicals causing mutations and cancer, *Science*, 204, 587, 1979.

Ames, B.N., Gold, L.S., and Willett, W.E., The causes and prevention of cancer, *Proc. Natl. Acad. Sci. USA*, 92, 5258, 1995.

European Economic Community, Genotoxicity: specific aspects of regulatory genotoxicity tests for pharmaceuticals. Legislative Basis Directive 75/318/EEC, Brussels, 1995.

International Agency for Research on Cancer, http://www.iarc.fr.

Tavtigian, S.V., Pierotti, M.A., and Borresen-Dale, A.L., International Agency for Research on Cancer Workshop on expression array analyses in breast cancer taxonomy, *Breast Cancer Res.*, 8, 303, 2006 [Epub ahead of print].

Weisburger, J.H. and Williams, G.M., The decision-point approach for systematic carcinogen testing, *Food Cosmet. Toxicol.*, 19, 561, 1981.

Williams, G.M. and Weisburger, J.H., Application of a cellular test battery in the decision point approach to carcinogen identification, *Mutat. Res.*, 205, 79, 1988.

REVIEW ARTICLES

Benigni, R. and Giuliani, A., Putting the predictive toxicology challenge into perspective: reflections on the results, *Bioinformatics*, 19, 1194, 2003.

Bennett, L.M. and Davis, B.J., Identification of mammary carcinogens in rodent bioassays, *Environ. Mol. Mutagen.*, 39, 150, 2002.

Bristol, D.W., Wachsman, J.T., and Greenwell, A., The NIEHS Predictive Toxicology Evaluation Project: chemcarcinogenicity bioassays, *Environ. Health Perspect.*, 104, 1001, 1996.

Carr, C.J. and Kolbye, A.C. Jr., A critique of the use of the maximum tolerated dose in bioassays to assess cancer risks from chemicals, *Reg. Toxicol. Pharmacol.*, 14, 78, 1991.

Chico-Galdo, V. et al., Acrylamide, an *in vivo* thyroid carcinogenic agent, induces DNA damage in rat thyroid cell lines and primary cultures, *Mol. Cell. Endocrinol.*, 257, 6, 2006.

Cimino, M.C., Comparative overview of current international strategies and guidelines for genetic toxicology testing for regulatory purposes, *Environ. Mol. Mutagen.*, 47, 362, 2006.

Cohen, S.M. and Ellwein, L.B., Cell proliferation in carcinogenesis, *Science*, 249, 1007, 1990.

Contrera, J. and DeGeorge, J., *In vivo* transgenic bioassays and assessment of the carcinogenic potential of pharmaceuticals, *Environ. Health Perspect.*, 106, 71, 1998.

Doe, J.E. et al., A tiered approach to systemic toxicity testing for agricultural chemical safety assessment, *Crit. Rev. Toxicol.*, 36, 37, 2006.

Duerksen-Hughes, P.J. et al., Induction as a genotoxic test for 25 chemicals undergoing in vivo carcinogenicity testing, *Environ. Health Perspect.*, 107, 805, 1999.

Gregory, A.R., Species comparisons in evaluating carcinogenicity in humans, *Reg. Toxicol. Pharmacol.*, 8, 160, 1988.

Helma, C. and Kramer, S., A survey of the predictive toxicology challenge 2000–2001, *Bioinformatics*, 19, 1179, 2003.

Ikeda, M. et al., Combined genotoxic effects of radiation and a tobacco-specific nitrosamine in the lungs of gpt-delta transgenic mice. *Mut. Res.*, 2006 [Epub ahead of print].

Jacobson-Kram, D., Sistare, F.D., and Jacobs, A.C., Use of transgenic mice in carcinogenicity hazard assessment, *Toxicol. Pathol.*, 32, 49, 2004.

Jones, S. and Kazlauskas, A., Growth-factor-dependent mitogenesis requires two distinct phases of signaling, *Nat. Cell Biol.*, 3, 165, 2001.

Kane, A.B., Animal models of malignant mesothelioma, *Inhal. Toxicol.*, 18, 1001, 2006.

Keenan, K.P., Laroque, P., and Dixit, R., Need for dietary control by caloric restriction in rodent toxicology and carcinogenicity studies, *J. Toxicol. Environ. Health B Crit. Rev.*, 1, 135, 1998.

MacDonald, J. et al., The utility of genetically modified mouse assays for identifying human carcinogens: a basic understanding and path forward, *Toxicol. Sci.*, 77, 188, 2004.

Mebust, M., Crawford-Brown, D., Hofmann, W., and Schollnberger, H., Testing extrapolation of a biologically based exposure-response model from *in vitro* to *in vivo* conditions. *Regul. Toxicol. Pharmacol.*, 35, 72, 2002.

Moolgavkar, S.H. and Knudson, A.G., Mutation and cancer: a model for human carcinogenesis, J. Natl. Cancer Inst., 66, 1037, 1981.

Perera, E.P., Perspectives on the risk assessment of nongenotoxic carcinogens and tumor promotors, *Environ. Health Perspect.*, 94, 231, 1991.

Rao, G.N. and Huff, J., Refinement of long-term toxicity and carcinogenicity studies, *Fundam. Appl. Toxicol.*, 15, 33, 1990.

Rao, P.M. et al., Dietary and metabolic manipulations of the carcinogenic process: role of nucleotide pool imbalances in carcinogenesis, *Toxicol. Pathol.*, 15, 190, 1987.

Roe, F.J., Refinement of long term toxicity and carcinogenesis studies, *Fundam. Appl. Toxicol.*, 16, 616, 1991.

Slamenova, D. Contemporary trends in *in vivo* and *in vitro* testing of chemical carcinogens, *Neoplasma*, 48, 425, 2001.

Tharappel, J.C. et al., Regulation of cell proliferation, apoptosis, and transcription factor activities during the promotion of liver carcinogenesis by polychlorinated biphenyls, *Toxicol. Appl. Pharmacol.*, 179, 172, 2002.

Waddell, W.J., Fukushima, S., and Williams, G.M., Concordance of thresholds for carcinogenicity of N-nitrosodiethylamine, *Arch. Toxicol.*, 80, 305, 2006.

Section III

Toxicology Testing In Vitro

11 Introduction to *In Vitro* Toxicology Testing

11.1 HISTORY OF *IN VITRO* METHODS

The understanding of the principles and mechanisms of the biomedical sciences blossomed significantly with the development of cell and tissue culture techniques. By the 1960s, the development of cell culture methods to grow animal and human cells and tissues on plastic showed promise in the survival of cells outside intact organisms. With further refinement of the culture conditions and insights into the requirements for maintaining viability of organs, tissues, and cells in an artificial environment, a transformation in cell biology was underway.

A variety of scientific disciplines began to appreciate the breakthroughs realized with these new techniques. Soon after that, in the 1970s, toxicologists adapted cell and tissue culture methodologies as pathways for revealing mechanistic toxicologies — the mechanisms of actions of toxicants. Cell culture broadened the scope of cell and cancer biology and other toxicologists came to realize that cells grown *in vitro* also retained the properties of their organs of origin. Perhaps these surrogate "tissues" grown in plastic could mimic animal tissues and organs in their responses to toxic insults.

The term *in vitro toxicology* was born and generally referred to the handling of tissues outside of intact organ systems under conditions that support their growth, differentiation, and stability. Since the seminal experiments established the validity of cell culture techniques* in toxicology, the methods have proven fundamental for understanding and developing critical procedures in cellular and molecular biology, as well as in pharmacology, genetics, reproductive biology, and oncology. Cell culture technology has improved dramatically as a result of interest in the methods and through its broad applications.

Although much of the mystery originally surrounding the techniques has waned, the doctrine underlying the establishment of the methodology remains. For example, improvements in our knowledge of serum requirements in media have fostered continued excitement in the roles of growth factors and cell differentiation. Without these techniques, the recent discoveries in the field of stem cell biology may not have been realized. Consequently, the discipline of cell culture and its application to *in vitro* toxicology and biology represent a tool poised to answer questions in the biomedical sciences that can only be addressed by examining the proliferation of

---

* The terms *cell culture* and *tissue culture* are used interchangeably in this section.

isolated cells without the influence of other organ systems. With this understanding, the use of cell culture does not purport to represent the whole human organism, but can significantly contribute to our understanding of the workings of its components.

11.2 *IN VITRO* SYSTEMS

The use of cell culture techniques in toxicological investigations is known as *in vitro cytotoxicology,* or *in vitro toxicology.* The latter term also includes non-cellular test systems, such as isolated organelles or high throughput (HTP) microarrays (see Part III, Chapter 17). The increasing interest and success of these techniques in toxicology are consequences of the considerable advancements realized since the advent of the technology:

1. Since the growth of the first mammalian cells in capillary glass tubes was described, the technology has progressed and has been refined extensively.
2. The mechanisms of toxicity of chemicals in humans and animals have been elucidated through the use of *in vitro* toxicological methods.
3. The necessity for determining the toxic effects of industrial chemicals and pharmaceuticals that are developed and marketed at rapid and unprecedented rates has made it compulsory to develop fast, simple, and effective alternative test systems.
4. Although the normal rate of progression of any scientific discipline is determined by progress within the scientific community, some areas have received more encouragement than others.

Specifically, the public objections and disapproval of animal testing have forced academic institutions, industrial concerns, and regulatory agencies to direct research initiatives toward the development of alternative methods of toxicity testing.

To understand the possibilities for and, more importantly, the limitations of cell culture methods in toxicology, it is necessary to become acquainted with the main features of the techniques for culturing cells and tissues. Unlike other types of biomedical research, these techniques require continuous care of cells in culture for extended periods and this necessitates planning.

11.3 HISTORY OF CELL CULTURE

11.3.1 Tissue Explants

In the fledgling years of cell culture, United States scientists removed tissue explants from animals and allowed them to adhere to glass cover slips or placed them in capillary tubes in clots formed from lymph or plasma. Figure 11.1 shows an early attempt to grow animal cells in glass tubes or plates following what appeared to be aseptic technique. The discovery was made early on that a tissue or organ that was once an intact biological specimen would exhibit a breakdown in the supporting matrix, followed by migration of individual cells from the specimen as a consequence.

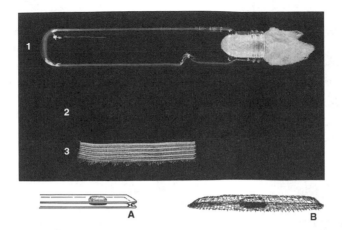

FIGURE 11.1 Scan of original diagram showing examples of glass (A) and fabric (B) envelopes and photographs of Leighton tubes (1) and envelopes (2 and 3) for the preparation of primary explants or dispersed animal cells (*Source:* from Robertson, A.L., *Tissue Cult. Assn. Man.*, 1, 167, 1975. With permission.)

With the addition of serum or whole blood, the resulting clot formed a hanging drop, making it possible to see these cells through an ordinary light microscope. With some refinement, the explants demonstrated that cells migrated out of a dissected specimen. This discovery made it possible to cultivate individual cells bathed in plasma or embryonic serum extract in small glass tubes. Aseptic techniques were further refined and the refinements allowed cells to be maintained in culture for longer periods. As the cells proliferated in higher quality tissue culture flasks, they could be transferred from one flask to another by scraping and loosening the growing monolayers with a rubber "policeman" (cell scraper).

Synthetic media (Earle, Parker, Eagle) were developed quickly and used with various serum additives. The main problem encountered was bacterial and fungal contamination which, because of the more rapid rate of mitosis of the microorganisms, usually outpaced the growth of the mammalian cells. This contamination generally resulted in overwhelming bacterial cell growth and disintegration of the cell cultures. This setback was largely overcome by the addition of liquid antibiotics to the media.

Later on, the development of better aseptic techniques such as the incorporation of sterile disposable glassware, autoclave units, and laminar air flow hoods made antibiotics superfluous in most cases. With an increase in the understanding of the influences of pH, buffers, gases, and ambient environments and the incorporation of chemically inert plastics and microprocessor-controlled incubators, came the realization of the full potential of *in vitro* technology.

11.3.2 Recent Developments

Today, cell biologists have further developed culture techniques to aid in their understanding of cellular and extracellular interactions such as mesenchymal–epithelial relationships, epithelium–cell matrix interactions, and stem cell biology. These *in vitro* studies have materialized largely through the development of additions to accepted protocols such as chemically defined cell culture media, the addition of cellular substrata to culture flasks, the introduction of porous membranes and filter inserts that allow for the passage of low molecular weight soluble substances, the development of treated plastic surfaces, and incubation of cells with co-cultures.

The following chapters highlight the principles, mechanisms, and current developments in the field of *in vitro* toxicology testing. In particular, the *in vitro* techniques described here correlate in principle with the *in vivo* toxicology testing techniques described in Part II and emphasize the importance of the former as alternative technologies for animal toxicity testing.

SUGGESTED READINGS

Barile, F.A., Introduction to In Vitro Cytotoxicology: Mechanisms and Methods, CRC Press, Boca Raton, FL, 1994.
Eagle, H., Media for animal cell culture, Tissue Cult. Assn. Man., 3, 517, 1977.
Freshney, R.I., Culture of Animal Cells: A Manual of Basic Technique, 4th ed., Wiley-Liss, New York, 2000.
McGarrity, G.J., Serum quality control, Tissue Cult. Assn. Man., 1, 167, 1975.
Robertson, A.L., Envelop technique for selective isolation of cells from multilayer organ cultures for metabolic studies, Tissue Cult. Assn. Man., 1, 139, 1975.
Weiss, L., Cell contact phenomena, in Advances in Tissue Culture, Waymouth, C., Ed., Williams & Wilkins, Baltimore, 1970, p. 48.

REVIEW ARTICLES

Abir, R., Nitke, S., Ben-Haroush, A., and Fisch, B., *In vitro* maturation of human primordial ovarian follicles: clinical significance, progress in mammals, and methods for growth evaluation, *Histol. Histopathol.*, 21, 887, 2006.
Aschner, M., Fitsanakis, V.A., dos Santos, A.P., Olivi, L., and Bressler, J.P., Blood–brain barrier and cell-cell interactions: methods for establishing *in vitro* models of the blood–brain barrier and transport measurements, *Methods Mol. Biol.*, 341, 1, 2006.
Rippon, H.J. and Bishop, A.E., Embryonic stem cells, *Cell Prolif.*, 37, 23, 2004.
Seth, G., Hossler, P., Yee, J.C., and Hu, W.S., Engineering cells for cell culture bioprocessing: physiological fundamentals, *Adv. Biochem. Eng. Biotechnol.*, 101, 119, 2006.
Turksen, K. and Troy, T.C., Human embryonic stem cells: isolation, maintenance, and differentiation, *Methods Mol. Biol.*, 331, 1, 2006.

12 Cell Culture Methodology

12.1 CELL CULTURE LABORATORY

12.1.1 EQUIPMENT

The cell culture laboratory is a sophisticated, meticulous, detail-oriented environment that relies on well-practiced features of cell biology while incorporating the latest instrumentation of biotechnology. The equipment, supplies, and analytical instruments are important aspects of the cell culture laboratory and are relied on for sensitive detection of cellular events. As the technology has advanced, so has the laboratory environment. Consequently, several important features of the cell culture laboratory are essential, including the following and other aspects of cell and tissue culture discussed in subsequent sections:

1. Sterile air flow work station
2. Temperature-controlled incubator
3. Autoclave
4. Cell counter
5. Custom gas tanks
6. Source of ultrapure water

Biological safety cabinets (Class II) reduce microbial contamination and also protect operators from exposure (Figure 12.1). These units are used for handling mammalian cells and also potentially infectious microorganisms, thus removing microbes within the sterile atmosphere while simultaneously protecting the operator from exposure. Alternatively, laminar flow hoods are unidirectional air flow vents that isolate the operating environment by blowing sterile air over the surface of the working platform or table top out toward the operator, rendering the operator vulnerable to exposure. This apparatus is generally reserved for preparation of sterile cell culture media, pharmaceutical injectables, and solutions.

Automatic glassware washing facilities, an incubator with digital humidity and temperature controls, and a dependable inverted microscope facilitate experiments and allow for less troublesome, routine operations in a cell culture laboratory (Figure 12.2). Thus, adequate tissue culture facilities fill a critical and dedicated role in toxicological investigations. In general, in order to maintain sterility, viability, identification, and integrity of cultured cells, the room containing the tissue culture

FIGURE 12.1 Biological safety cabinet: 5-ft purifier digital series class II, type A2. (Photo courtesy of Labconco Corporation, Kansas City, Missouri.)

FIGURE 12.2 Air-jacketed automatic CO_2 incubator. (Photo courtesy of Barnstead/Thermolyne, Barnstead International, Dubuque, Iowa.)

facilities is dedicated to this function. However, complete separation of all functions is not always possible or necessary. For instance, an analytical balance, pH meter, hot plate, and magnetic stirrer can be shared with other laboratory areas, whereas media preparation facilities, cell culture equipment, and glassware are generally reserved for the cell culture laboratory.

Most supplies and plastic ware used in handling cell cultures routinely are sterile and disposable. These include borosilicate glass or plastic pipettes and bottles, polyethylene tissue culture flasks, petri dishes, polypropylene and polyethylene centrifuge tubes, and filtering supplies. The disposable supplies reduce microbiological contamination and integration of other cultured cells present in the laboratory. In addition, sterile, ready-formulated media may be purchased and save the inconvenience of preparation time.

12.2 CULTURED CELLS

12.2.1 PRIMARY CULTURES

In general, establishing cells in culture from intact, viable tissue explants* is easily facilitated from younger donors, that is, the younger the age of the mammalian or human donor, the more replications and shorter replication time are expected from the resultant cells. In addition, cells from embryonic tissue are quickly established in culture, usually progressing through more population doublings *in vitro* than cells from adult donors. Some cells derived from neoplastic tissue are similar to embryonic cells in this respect.

Ideally, cell cultivation begins with the aseptic removal of an explant ($1/2 \times 1$ mm) from a tissue or organ specimen. The cells of the explant are mechanically

---

* Explant cultures are viable anatomical specimens removed from a mammalian donor and prepared for cell culture, usually 1 to 3 mm in diameter. They have the capability of generating primary cultures.

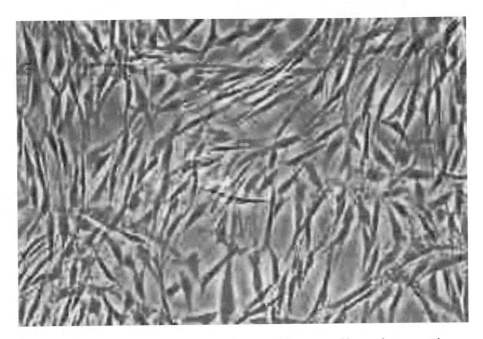

FIGURE 12.3 Human lung fibroblasts (HLFs) derived from normal human lung parenchyma. (Photo courtesy of Cell Applications, Inc., San Diego, California.)

and/or enzymatically separated from the matrix and are allowed to grow in medium in contact with the bottom surface of the culture vessel. While the central core of the explant often atrophies due to less favorable diffusion of nutrients through the tissue, peripheral cells migrate outward and proliferate. The cells migrating from the tissue establish a *primary culture*. Two important and parallel processes occur during this transformation:

1. The differentiated cells of the original explant may divide, depending on the organ of origin; also, as the culture progresses, the established primary cells successively lose some of their specialized functions — a process referred to as dedifferentiation.
2. Less specialized cells such as mesenchymal cells (e.g., fibroblasts) divide rapidly, and eventually outpace the specialized cells (Figure 12.3).

12.2.2 FINITE CONTINUOUS CELL LINES

As the cells from the primary culture proliferate, they eventually occupy the entire vessel surface exposed to the medium. Thus, the culture reaches a state of *confluency* and microscopically appears as a uniform, closely attached carpet of cells known as a *monolayer*. The cells are ready for transfer to new culture vessels containing fresh medium. With this manipulation, the culture ceases to be in a primary state and is designated a *continuous cell line*. Subcultivation, or *passaging*, is achieved by detaching the cells from the glass or plastic using a calcium chelator solution such as EDTA or by exposing the monolayer surface to enzymatic methods.*

The cells are dispersed, collected, centrifuged, resuspended, counted, and inoculated in several new vessels. Because of the selective survival of viable cells, the cell line is more homogeneous and dedifferentiates with time. A culture that enjoys adequate proliferation is harvested as above every 3 to 7 days, depending on the cells, and subdivided to a new environment.

Cell lines established from non-neoplastic tissue have finite aging processes *in vitro* and ultimately stop dividing. They exhibit a specific number of cell divisions — usually determined experimentally.† Such a *finite cell line* has a constant (diploid) number of chromosomes and exhibits an orderly oriented growth pattern including inhibition of growth of individual cells by contacts with their neighboring cells (*contact inhibition*).

It is possible to select a single cell from a confluent primary culture and transfer it to a culture flask. The proliferation of cells from a single primary cell is designated a *clonal cell line*. Cell culture cloning, according to the traditional definition, allows for the selection of one type of cell (e.g., a cell with a specific functional marker) and the establishment of a subculture in which the desired characteristic has been passed on to the progeny.

12.2.3 IMMORTAL CONTINUOUS CELL LINES

Cell lines that do not age in culture with continuous passage are transformed to *immortal continuous cell lines*, principally because their growth patterns do not exhibit typical signs of aging in culture. Transformation occurs either spontaneously or is induced as a result of incubation with viruses or carcinogenic or mutagenic chemicals.

The exact mechanism of the transformation involves the initiation and promotion of stages of carcinogenesis. Thus a continuous cell line acquires a set of characteristics such as a varied chromosome number (heteroploidy) and loss of contact inhibition. Moreover, immortal continuous cell lines are able to form colonies in soft agar media — that is, without help of glass or plastic contact — and induce tumors if implanted in immunologically nude animals. In spite of this, some highly differentiated cell functions that mimic those of specialized cells persist in immortal cell lines.

These differentiated functional markers are used to characterize the cells and are relied upon to monitor the progress of the monolayers in culture, particularly because the low power morphological appearances of finite and immortal cells *in vitro* are not easily distinguishable. The criteria are also used as parameters for assessing *in vitro* toxicity. Table 12.1 presents some frequently used indicators of toxicity as general markers of cell viability and function.

* Because the enzyme of choice is trypsin, the process of subcultivation of monolayers is often referred to as trypsinization.
† The finite replications are also referred to as a population doubling level (PDL) and may require months in culture, depending on growth rate.

TABLE 12.1
General Viability Toxicity Criteria for Cell Lines

Indicator	Methods for Determining Degree of Toxicity
Cell morphology	Histological, ultrastructural, immunohistochemical analysis
Cell proliferation	Cell count, mitotic frequency, DNA synthesis, karyotypic analysis
Cell division	Plating efficiency, clonal formation
Cell metabolism	Uptake of fluorescent or isotope-labeled precursors, fluorescent luminescence assays for enzymatic activity
Cell membrane	Leakage of enzymes (LDH), uptake of trypan blue
Cell staining	Immunohistochemical or cytochemical stains for cell markers
Cell differentiation	PCR analysis for gene markers
Mitochondria	Mitochondrial reduction (MTT assay)
Lysosomes	Vital staining, NRU assay

12.2.4 STEM CELL LINES

Embryonic stem (ES) cells are derived from pluripotent cells of the early mammalian embryo and are capable of unlimited, undifferentiated proliferation *in vitro*. In chimeras with intact embryos, ES cells contribute to a wide range of adult tissues including germ cells, providing a powerful approach for introducing specific genetic changes into a cell line. Pluripotency allows cells to differentiate into many cell types and tissues, including all three embryonic germ layers. Furthermore, experimental evidence has shown that organogenesis of embryonic and adult stem cells to neural, muscle, dermal, and bone marrow lineages is possible. Figure 12.4 illustrates a general outline of the steps involved in the derivation of differentiated ES cells in culture.

The concept of deriving ES cells from mice and humans is relatively recent. The protocols developed to maintain ES cells in continuous culture continue to emerge. Some progress, however, has demonstrated the transformation of ES cells to different organ and tissue structures including *in vitro* organogenesis and differentiation of human, primate, and mouse ES cells into functional gut-like organs, smooth muscle cells, cardiomyocytes, neurons, and hematopoietic entities. In addition, soluble factors influence direct differentiation of mouse ES cells. For example, IL-3 directs cells toward macrophage, mast cell, neutrophil, or erythroid lineages. Retinoic acid induces neuron formation, and transforming growth factor (TGF) induces myogenesis. Differentiation is also encouraged by culturing ES cells on different feeder layers and extracellular matrix components that exhibit morphological and physiological properties characteristic of the target organ.

The increasing awareness of the influence of growth factors and basement membrane substrata has improved our understanding of the role of cell regeneration in wound repair and in pre-neoplastic tumor growth. In their ability to form tight junction strands, differentiating epidermal and epithelial cells act to prevent free interchange of most solutes between luminal and interstitial fluids along the paracellular route. Breakdowns in the barrier functions of these membranes may con-

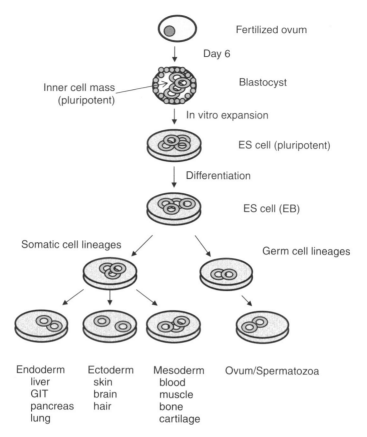

FIGURE 12.4 Embryonic stem (ES) cells isolated from the inner cell mass of a blastocyst. Under the influence of different growth factors, ES cells have the potential to differentiate to cells of different lineages (EB = embryoid bodies).

tribute to cellular neoplasia. Thus, continuous progress in stem cell biology and biotechnology is based on the understanding that *in vivo* epithelial, epidermal, and mesenchymal cells undergo continuous renewal by multipotent stem cells that remain anchored to the organ basement membrane. Thus a single stem cell migrates and commits to all of the embryonic lineages.

12.2.4.1   Influence of Culture Environment on ES cells

When ES cells are allowed to differentiate in suspension culture or in the absence of feeder layers, they form spherical multicellular aggregates or embryoid bodies (EBs) that contain a variety of cell types. By manipulating combinations of growth factors (GFs) in the presence of extracellular matrix (ECM) components, epithelial- or epidermal-specific gene expression and differentiation are induced.

The mechanism by which GFs and ECM influence cell differentiation or expression of differentiation markers is the subject of much interest. In the presence of GFs in culture, mouse ES cells acquire different expression profiles and display evidence for tight junctions. It is understood that without GFs, leukemic inhibitory factor (LIF), or feeder layers, stem cells spontaneously differentiate into a variety of uncontrolled cell colonies and properties for cellular differentiation are delayed.

Thus, GFs and ECM are capable of directing linear specific differentiation. For instance, epidermal growth factor (EGF) is a mitogenic polypeptide that allows for differentiation into ectoderm (including skin) and mesoderm, while keratinocyte growth factor (KGF) is an epithelial cell-specific mitogen responsible for normal proliferation and differentiation of epithelial cells. Consequently, differentiation of ES cells in culture to specific lineages may be mediated by enrichment of the environment with GFs that influence either cell expansion, improved survival, or differentiation.

12.2.4.2 Culture of Human or Mouse ES cells

Currently, most experimental stem cells are derived from blastocysts of humans or mice. The cells spontaneously differentiate into embryonic structures in the absence of a feeder layer or conditioned medium. They are maintained in the undifferentiated state by frequent subculture (every 2 days) on confluent mitomycin-C treated or irradiated mouse embryonic feeder layers (MEFs) or in complete culture medium to which is added leukemic inhibitory factor (LIF), or both.*

In Figure 12.5, undifferentiated mouse ES cultures are distinguished from differentiated populations and from the underlying MEFs by the formation of EB. These aggregates are four to ten cell layers thick over the feeder layer, are oval or round, and have defined borders. As undifferentiated mouse ES cell colonies propagate, they maintain their aggregate morphology (small arrow, Figure 12.6) and are still distinguished from the feeder layer (asterisk). Figure 12.7 shows differentiated

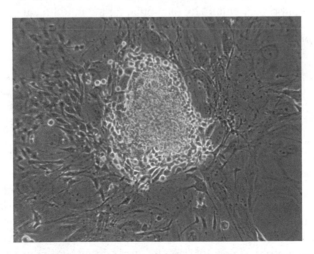

FIGURE 12.5 Phase contrast image of embryoid bodies (EBs) from mouse ES-D3 cells after removal of LIF. Aggregates are round and have well defined borders. They maintain their aggregate morphology and are distinguished from the underlying feeder layer (400×).

---

* Recombinant human LIF is a lymphoid factor that promotes long-term maintenance of mouse ES cells by suppressing spontaneous differentiation; it is not effective in human ES cells.

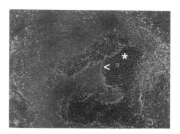

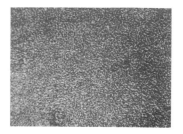

FIGURE 12.6 In the absence of feeder layers and LIF, undifferentiated EBs from mouse ES cell colonies propagate, lose their aggregate morphology, and migrate to the periphery as they continue to differentiate.

FIGURE 12.7 Differentiated ES cells in the absence of MEF or LIF.

ES cultures when cells are allowed to come into contact, forming confluent monolayers — a transformation that occurs in the absence of feeder layers and LIF. The figure features the loss of aggregate morphology by ES cells.

Protocols for the preparation of mitomycin-C-treated MEF feeder layers, culture media, serum requirements, and methods for subculturing and maintenance in undifferentiated or differentiated states, are described extensively in selected Suggested Readings and Review Articles at the end of this chapter.

12.2.5 CLONAL GROWTH AND MAINTENANCE CULTURES

Highly differentiated finite cell lines from explanted or enzymatically treated tissue are maintained in primary culture by manipulating the liquid medium environment. Such techniques include the use of serum-free medium, arginine-free medium, and coating of the plastic culture surfaces with ECM (laminin or fibronectin) that promotes growth and adhesion of either epithelial or mesenchymal cells, respectively. In addition, cell types are mechanically separated by employing *clonal growth* or *differential adhesion* manipulations — in whole tissue, epithelial cells are separated from fibroblasts based on the ability of the latter cells to settle and adhere to the surface more quickly. One to three hours after an initial inoculation of the culture with the cell suspension, as determined empirically, the medium is removed and the epithelial cells that have not adhered are plated in a separate vessel. The primary culture now serves as a *maintenance culture* of specialized cells.

Proliferation and growth of mortal maintenance cultures depends on several factors including organ of origin, age of donor, and cell type. For instance, normal primary cultures of specialized cells, including those derived from adult rodent liver, may be kept for weeks, while cells derived from embryonic liver, neuronal, or lung epithelium are viable for months. An established culture always has a tendency to lose specific functions with time, but this process may be counteracted by culturing the specialized cells on MEF feeder layers or by using serum-free medium. Isolation of primary cultures, however, requires repeated use of animals for extraction and establishment of explants, the costs of which are described in Chapters 6 and 7 in Part II.

12.2.6 CRITERIA FOR IDENTIFICATION AND MONITORING OF CULTURED CELLS

Various criteria, cell markers, and indicators are used for the identification and classification of cells in culture. These methods are employed when a cell line is established and periodically thereafter to monitor the genetic purity of the cells. The criteria are described below.

12.2.6.1 Karyotypic Analysis

When a designated cell line is derived from normal tissue, the chromosome complement should be identical to the parent cell or the species of origin. In this way, the cell line is classified as diploid (2n chromosomes). Any other designation would classify the cells as aneuploid or heteroploid. Continuous cell lines derived from tumors or transformed cells fall into the latter categories.

12.2.6.2 Aging in Culture

The life spans of cells in culture are measured according to the population doubling level (*pdl*). This calculation monitors the frequency of mitotic events characteristic of proliferating cells after their original establishment in primary culture. The general formula for assessing *pdl* is:

$$pdl_f = 3.32 \, (\log F - \log I) + pdl_i$$

where pdl_f represents the final *pdl* at the time of trypsinization or at the end of a given subculture; F is the final cell count; I is the initial cell number used to initiate the individual culture vessel at the beginning of the subculture; and pdl_i equals the doubling level of the cells used to initiate the subculture.

Cells with finite life spans generally show signs of aging such as loss of cell morphology, decrease in rate of proliferation, increase in cytoplasmic lipid content, and other evidence of apoptosis. The amount of *pdl* exhibited by any particular finite cell line or primary culture depends largely on the age of the organ of origin. Thus the *pdl* is ascertained from the literature or determined empirically. Immortal continuous cell lines are capable of indefinite multiplication *in vitro*, provided they are maintained under optimum conditions. Monitoring of the *pdl* is not always necessary with immortal cell lines.

12.2.6.3 Anchorage-Dependent Cultures

Some cells are capable of growing in suspension culture, while others require attachment to plastic. This plastic support may take the form of polyethylene or polypropylene treated so that its electrostatic properties allow isolated cells to attach and migrate. Other attachment matrices include components of the ECM such as laminin and fibronectin. These components mimic the *in vivo* attachment substrata normally present in the epithelial lining and the interstitium, respectively. Thus, they may be selective for cells of epithelial or mesenchymal origin. Still other support matrices use microporous filter membranes that both allow for cellular attachment and also permit passage of macromolecules through the membranes to the basolateral surface of the cell layer.

12.2.6.4 Contact Inhibition

As mortal continuous cells proliferate in a sparse culture environment, they migrate and occupy the available surface that is inundated with medium. As the surface becomes congested, the cells necessarily come into cell membrane-to-membrane contact. Although the monolayer is viable, proliferation eventually ceases and the cells exhibit *contact inhibition* characterized by arrest of the cell cycle in the G_0 phase. Figure 12.8 illustrates the cell cycle exhibited by proliferating cells *in vitro* or *in vivo*.

Cells that do not exhibit contact inhibition grow in multilayers, successively loosening attached cells from the support as space for growth and proliferation becomes competitive. Consequently, although confluent multilayer cultures do not exhibit contact inhibition, cells are often discovered floating in the surrounding medium. Most continuous immortal cell lines do not succumb to growth restrictions demonstrated by mortal cells.

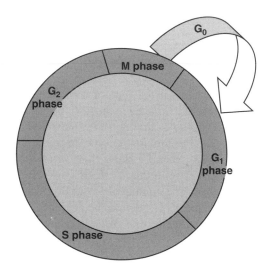

FIGURE 12.8 Four successive phases of cell cycle for proliferating cells *in vitro* or *in vivo*. Mortal proliferating somatic cells spend most of their existence in interphase. As a cell prepares to divide, it enters the G_1 *phase* lasting from several hours to days and characterized by synthesis of cell organelles and centriole replication (the G_0 *phase* represents a resting stage or subphase of G_1). The *S phase* begins when DNA synthesis starts and the chromosomes have replicated. The G_2 *phase* starts when DNA replication is complete and the content of the nucleus has doubled. The extended part of interphase then continues in the *M phase* with the stages of cell division: prophase, metaphase, anaphase, and telophase. The *M phase* begins with mitosis (nuclear division) and ends with cytokinesis (cytoplasmic division). Cell division terminates with the completion of cytokinesis.

Other methods of cell classification are not customarily performed but are incorporated into mechanistic experiments. These include analysis of tissue-specific differentiated properties such as secretion of macromolecules or the presence of intracellular enzymes, ultrastructural identification of structure–function activity, fluorescent labeling for identifiable structural proteins, RT-PCR (reverse transcriptase polymerase chain reaction) analysis for expression of gene-specific primers and the ability of the cells to form an invasive malignant tumor when injected into immunologically deficient (nude) mice.

12.3 CULTURE REQUIREMENTS

12.3.1 MEDIA COMPONENTS

Cells are cultured in a liquid medium whose components are precisely controlled and monitored. The medium is composed of a buffered solution of physiologic ion concentration containing soluble amino acids, carbohydrates, vitamins, minerals, fatty acids, and other cofactors. Optional ingredients include a pH indicator, separate buffering systems such as HEPES, and some non-essential amino acids that are incorporated when required by a particular cell type. The formulas for the media are readily available (see websites listed in Suggested Readings) in soluble powdered or liquid formulas.

Some of the more commonly used media are modified Eagle's medium (MEM), basal medium Eagle (BME), Dulbecco's modified Eagle's medium (DMEM), and Ham's F12 medium. These solutions are generally designed for use with serum or serum proteins with the addition of fetal or newborn mammalian serum as needed.

Serum is a complex mixture of poorly defined biological components consisting of soluble cell growth factors, transport proteins, hormones, essential trace metals, lipids, and ECM adhesion factors. Without the addition of a minimum percentage of serum (5 to 20%) as part of the formulation, cellular proliferation does not proceed favorably.

In cell culture, balanced salt solutions function as washing, irrigating, transporting, and diluting fluids while maintaining intracellular and extracellular osmotic balance. In addition, the solutions provide cells with water and inorganic ions essential for normal cell metabolism and are formulated to provide a buffering system to maintain the liquid within the physiological pH range (7.2 to 7.6). The solutions are also combined with carbohydrates such as glucose, thus supplying an energy source for cell metabolism during extensive wash procedures. The most commonly used prepared salt solutions include Dulbecco's phosphate buffered saline (PBS), Earle's balanced salts, and Hanks' balanced salt solution (HBSS).

In addition to satisfying nutrient requirements, the cells must be placed in an environment that mimics the *in vivo* situation. Table 12.2 summarizes the minimum conditions that support survival, maintenance, and proliferation of cells in culture. The parameters include temperature, pH, carbon dioxide tension, buffering, osmolality, humidification, and the absence of microbial contamination (sterile technique). They vary only slightly, depending on the cell type.

TABLE 12.2
General Minimum Requirements for Maintenance and Proliferation of Cultured Cells

Requirement	Parameters
Culture conditions	pH, buffered growth media and wash solutions, temperature, humidity, osmolality, ultrapure water
Technical	Sterile technique, patience
Equipment	Biological safety cabinet or laminar flow hood, CO_2 incubator, autoclave, water purification system
In vitro medium supplementation	FBS, FCS, NCS, NHS, ECM components, MEFs, culture inserts, defined media additives, pretreated culture vessels
Supplies	Plastic culture wells, plates, flasks, and other disposable supplies
Optional components	pH indicators, HEPES buffer, NEAA, antibiotics, co-cultures

Note: ECM = extracellular matrix. FBS = fetal bovine serum. FCS = fetal calf serum. MEFs = mouse embryonic feeder cells. NCS = newborn calf serum. NEAA = non-essential amino acids. NHS = newborn horse serum.

12.3.2 TEMPERATURE

For most mammalian cells, the optimum temperature for cell proliferation and differentiation is 37°C. Cells withstand falls in temperature more easily than rises. In fact, temperatures as low as 4°C may reduce metabolic activity but will not irreversibly inhibit biological functions. Temperatures as high as 39°C may stimulate the formation of heat shock proteins, resulting in irreversible functional and ultra-structural damage and apoptotic events.

12.3.3 pH

The pH range that supports optimum cellular multiplication is 7.2 to 7.4, with some cell growth occurring at the extremes of 7.0 and 7.6. Some cells are more tolerant of slightly acidic conditions than corresponding basic conditions, and transformed cells are often more resistant to changes in pH than diploid cells. In addition, rapidly proliferating cells with high metabolic requirements eliminate excessive acidic waste material more quickly, resulting in rapid falls in pH even in the presence of strong buffers. These types of cells often require more frequent media washes. In general, pH is readily monitored in an incubator as well as during routine manipulation with the addition of pH indicators (e.g., phenol red) in the growth media and salt solutions. The indicators do not interfere with culture stability.

12.3.4 CARBON DIOXIDE TENSION

Most culture media contain bicarbonate as part of the buffering system to prevent large and rapid changes in pH. At room temperature and standard incubator pres-sures, bicarbonate and carbonic acid are in equilibrium in solution at physiologic pH. At 37°C, however, soluble carbon dioxide (CO_2) is driven out of solution, thus disrupting the equilibrium established among bicarbonate, carbonic acid, and CO_2. This condition disturbs the critical balance in the concentrations of carbonic acid and bicarbonate in the culture medium.

Thus, balanced CO_2 tension and maintenance of pH are satisfied by maintaining an increased partial pressure of CO_2 in the gas phase above the liquid. Modern water-jacketed CO_2 incubators are designed to adequately maintain the gaseous CO_2 phase in a chamber at controlled levels. Therefore, most culture media are formulated for use with 2 to 15% CO_2 so that the appropriate pH is maintained throughout the life cycles of the cells.

12.3.5 BUFFERING

When petri dishes, flasks, or multiwell plates are removed from an incubator chamber, CO_2 necessarily escapes from the chamber and from the warm liquid medium. This inevitably results in the inability of pH to be adequately maintained in the absence of the gas phase. This occurrence is not usually a problem when cell cultures are removed during washing steps with physiological salt solutions or during replacement of expended medium. When *in vitro* experiments necessitate long periods outside an incubator, however, a buffering system is required. Many laboratories routinely incorporate organic buffers such as HEPES in media formulations to prevent the rapid shift of pH when cultures are removed from a CO_2 incubator. Additionally, sufficient sodium hydroxide or sodium bicarbonate is added to reach the working pH.

When a gas or vapor environment must be maintained without interruption, as in experiments using gaseous toxins, cells are grown in screw-capped tissue culture flasks. The flasks are then perfused with specially formulated custom gas mixtures separate from the incubator environment and sealed with screw caps, thus preventing escape of the gaseous phase. Although tedious, the method is useful when analyzing the effects of volatile organic liquids or gases in toxicology testing protocols.

12.3.6 OSMOLALITY

The optimum range of osmolality in a culture medium depends on the cell type, but it is generally between 250 and 325 milliosmoles per kilogram (mOsm/kg). The osmolality of a solution is standardized during the preparation of the powdered medium. Concerns arise during the time of incubation when the relative humidity in the chamber is *not* kept at or near saturation. When an incubator door is opened, dry air enters and cools the chamber, thus releasing humidified air by evaporation. An incubator that does not provide for adequate heat transfer will remain cool. If the liquid media in the cultures is at higher temperature, the water in the media evaporates, leaving a solution with a higher osmolality.

To prevent evaporation and to maintain the proper osmolality of a culture medium, rapid equilibration of heat throughout the chamber is necessary. Modern water-jacketed incubators are equipped with internal humidity control devices and fans that distribute heat. These chambers are precisely insulated, heated, and humidified to prevent the formation of cool spots that may prevent the humidification of the chamber. Alternatively, the chamber humidity is maintained by leaving an open trough of water (preferably with a large surface area) at the bottom of the instrument. In older generation incubators, adequate humidification is determined visually by the presence of condensation on the glass door of the incubator, but no condensation should be formed on the culture dishes. Some incubators are equipped with heated inner doors that prevent condensation.

Humidification is also monitored by measuring the osmolality of control media incubated in parallel to cell cultures. With evaporation, the osmolality increases, as compared with fresh medium at the start of a cell cycle, indicating that the humidity is inadequate.

12.3.7 WATER REQUIREMENT

A fundamental requirement of cell culture systems, and one that is often overlooked in the initial steps of establishing a cell culture laboratory, is the quality of the water used to prepare media and salt solutions. Ordinarily, distilled water or a single passage of water through a deionizing column is not sufficient to remove all the impurities that remain from feed water. Contaminants in the water may take the form of trace metals, organics, variable amounts of divalent cations such as magnesium or calcium, and metabolic products of microorganisms. Such contaminants interfere with cell growth and functional processes.

The impurities are essentially removed with systems that recirculate water through a series of deionizing and organic exchange columns. Most of these systems are equipped with digital sensing monitors that display the purity of the collected water in megaohms (mega-Ω). As ions and impurities are removed in the filtration columns, water resistance increases. A resistance measuring at least 10 mega-Ω is suitable for culturing cells; the limit of purification with many ion exchange columns is about 18 mega-Ω. Some water purifying systems also yield water which is sterilized by gamma irradiation at the collecting end.

12.3.8 GLASSWARE REQUIREMENT

Another aspect of maintaining a cell culture laboratory at optimum efficiency is the manner in which the glassware is handled, washed, and stored. All glassware designated for use in a cell culture area should not be used for storage of other chemicals or for experiments unrelated to the growth of cells. Glassware is clearly marked *for cell or tissue culture only* and is washed in detergent compatible with the objectives of the laboratory. Essentially, the detergent should leave no residue adhering to the pores of the glass. It should rinse easily with deionized water. The glassware is also stored separately from other equipment and materials kept in the laboratory.

12.3.9 SERUM-FREE MEDIUM

In investigations dedicated to understanding the influences of nanomaterials on growth and proliferation, mechanisms of toxicity, and actions of growth factors or hormones on cultured cells, the use of batch fetal or newborn serum is not acceptable.

Serum is a complex and poorly characterized mixture, the composition of which varies according to the commercial lot. In addition, serum contains naturally occurring substances or microbiological contaminants (e.g., mycoplasma, viruses, endotoxins, prions) that may prove toxic or induce genetic or structural transformations for most cells.

Thus, the use of serum-free medium replaces serum in standard growth solutions by incorporating a mixture of synthetic or otherwise defined factors that substitute for essential culture requirements. The factors include hormones (insulin, transferrin), non-essential amino acids, trace metals (selenium, manganese, zinc), growth factors (epidermal growth factor, fibroblast growth factor, keratinocyte growth factor), and extracellular matrix components (ECMs such as iaminin and fibronectin). Selective serum-free media facilitate adaptation of cells to the culture environment and allow for better standardization of experimental conditions.

A serum-free medium also facilitates the isolation of the desired cell type during procedures for the establishment of primary cell cultures. Incorporation of serum-free medium as a culture requirement essentially eliminates the overgrowth of fibroblasts that usually proliferate rapidly in serum-supplemented media. Serum-free medium is individually prepared and adapted to each cell type. The advantages of integrating serum-free (or even low serum) media are well documented and have been described for use in maintenance cultures of many primary and specialized cells and for finite and continuous cell lines (see Suggested Readings and Review Articles at the end of this chapter).

12.4 PROCEDURES

12.4.1 STATIC AND PERFUSION CELL SYSTEMS

In general, culture conditions during an experiment or assay differ from those used during routine growth and maintenance of stock cultures. All stock cultures and most experiments are *static cell systems* that entertain a change of cell medium before or during subcultivation. During normal incubation, depletion of nutrients, production of metabolic acids, and accumulation of excretion products occur in the medium, resulting in less favorable cell culture conditions.

Cell types that are anchorage-dependent are cultured as monolayers on the bottom surface of a flask and covered with medium in addition to an overlying gas phase consisting of 2 to 15% CO_2 in air. Cells may also be anchored on filter inserts or polystyrene particles. Rotating flasks or tubes are used to increase surface area by providing contact of the sides and tops of the vessel with media, which also permits better oxygenation of cells. Some cell types (continuous cell lines) need not adhere to a substratum and may be cultured in magnetically stirred suspension cultures or roller bottles.

Perfusion cultures are more advanced compared to static cell cultures because they allow for the uniform flow of fresh medium by continuous irrigation of the flask. These systems also use positive pressure pumps and micron filters to provide the cultures with oxygenated, waste-free medium. Although maintenance of these systems is very fastidious, they may present some distinct advantages in metabolic toxicity testing studies. Note that most cell culture test systems are much less oxygenated than animal tissues and thus predominantly permit anaerobic cell metabolism.

12.4.2 Dispersion of Tissues

Many methods exist for the establishment of cell suspensions. The choice of protocol largely depends on the cell type desired and the experience of the investigator. Most of the commonly used procedures for isolating primary cultures or continuous cell lines employ the same principles underlying tissue dispersion. Briefly, the tissue is surgically extirpated and the cells are dispersed mechanically, chemically, and/or enzymatically.

Ordinarily, a typical protocol will use all these manipulations. A tissue is mechanically dissected, minced, or sheared and passed through nylon mesh or sterile gauze. Initial dispersion of tissue is further completed with repetitive pipetting, magnetic stirring, or gentle vortexing. Chemical dispersion involves the removal of calcium from the dispersant solution by the addition of chelators such as ethylenediamine tetraacetic acid (EDTA). Enzymatic dispersion requires the use of trypsin, chymotrypsin, collagenase, deoxyribonuclease (DNAse), or pronase. Figure 12.9 illustrates a typical protocol for isolating cells from animals and outlines preparation of the animal for anesthesia and surgical manipulation, general procurement of the organ, preparation and mechanical treatment of the tissue, dispersal in an enzymatic solution, and final recovery and plating of the suspended cells in growth medium.

In general, this procedure is typical for isolating cells from any intact visceral organ or when obtaining biopsy specimens, and should be performed in a biological safety cabinet. Explants are also established by placing individual dissected tissues in a drop of medium and allowing the specimens to adhere. Maintenance of aseptic conditions is essential for the establishment of successful cell cultures.

12.4.3 Subculturing

Subculturing or passaging of cells involves the mechanical or enzymatic disruption of the cell monolayer, and dividing and transferring the cells to new culture vessels. The procedure is required when a cell population has occupied most or all of the available surface area in a vessel. At this density level, as explained above, confluent normal diploid cells exhibit *contact inhibition* cell proliferation and migration slow or cease as cells come in contact with each other. Common methods used for the dispersion of cells include mechanical and enzymatic disruption.

Mechanical disruption involves the use of sterile silicone or plastic cell scrapers to dislodge the monolayer in aggregates. The method is not satisfactory for maintaining continuous cultures whose cell membranes are sensitive to physical manipulation because the process often compresses or lyses cells. The technique, however, is often used when terminating experiments for preparatory analyses.

Enzymatic disruption uses proteolytic enzymes such as trypsin, collagenase, and/or DNAses that yield satisfactory cell numbers and afford less cell destruction.

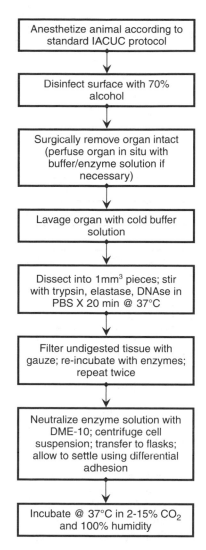

FIGURE 12.9 Typical procedure for mechanical and enzymatic isolation of cells from intact animals. All procedures are performed at 4°C using aseptic techniques. PBS = Dulbecco's phosphate buffered saline. DME-10 = Dulbecco's modified Eagle's medium containing 10% fetal bovine serum (see Section 12.2.5).

The enzyme solution is formulated as 0.5 to 2.0% in Ca^{+2}–Mg^{+2} free balanced salt solution (PBS) and supplemented with a cation chelator (EDTA). The cells are not left in contact with the enzyme for more than the prescribed time (about 10 min) because enzymes are capable of damaging cell membranes. Termination of enzyme activity is accomplished by neutralizing the activity of the solution by the addition of fresh medium containing serum. Figure 12.10 highlights the protocol generally used for enzymatic dispersion of cell monolayers.

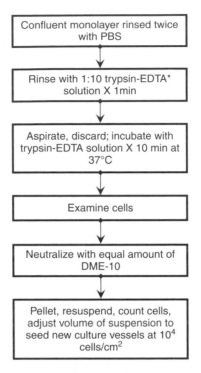

Confluent monolayer rinsed twice
with PBS

Rinse with 1:10 trypsin-EDTA*
solution X 1min

Aspirate, discard; incubate with
trypsin-EDTA solution X 10 min at
37°C

Examine cells

Neutralize with equal amount of
DME-10

Pellet, resuspend, count cells,
adjust volume of suspension to
seed new culture vessels at 10^4
cells/cm^2

* ETDA = ethylenediaminetetraacetic acid

FIGURE 12.10 Outline of general protocol used for enzymatic dispersion of cell monolayers. DME-10 = Dulbecco's modified Eagle's medium containing 10% fetal bovine serum. PBS = phosphate buffered saline. Asterisk = 1:10 dilution of 10X trypsin–EDTA concentrate in PBS.

After counting, the cell suspension is diluted and the cells are inoculated (seeded) into appropriate culture vessels. In most cases, cells are seeded at 10^4 cells/cm^2 of surface area, a seeding density corresponding to about one fourth to one third the confluent density, thus allowing for optimum cell proliferation. The newly seeded culture plates, wells, or flasks are incubated undisturbed for several hours to facilitate cell attachment.

12.4.4 MEASUREMENT OF GROWTH AND VIABILITY

Some important considerations when determining the integrity of a cell line or newly established primary cultures rest upon the knowledge and experience with cell culture in general and the cell type in particular. Overall, several criteria are assessed to determine the potential utility of the cells for future toxicology investigations including:

1. The viability of the cultures and appearance of the cells under a phase contrast microscope, especially as a finite cell line approaches the anticipated maximum number of doublings.
2. The ability of the cells to retain as many of the morphological and biochemical features as the original parent cell or primary culture. Most

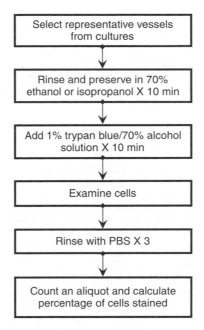

FIGURE 12.11 Outline of protocol using trypan blue vital dye exclusion method. PBS = phosphate buffered saline.

of these features are discussed in subsequent chapters since toxicological evaluations depend on the retention of these features.

More common criteria used to monitor cell proliferation and viability routinely include cell counting (as described above); the exclusion of the vital trypan blue dye; the ability to survive or proliferate under normal culture conditions; and the percentage of cells inoculated into the culture vessel that will attach to the culture surface. Each procedure is performed as part of the routine monitoring of culture status. Most of these features are determined under the phase contrast microscope, assuming that the individual responsible for the cultures has a familiarity with the cells in question. Figure 12.11 outlines the method of vital dye exclusion.

12.4.5 CRYOPRESERVATION

Most cell lines are frozen indefinitely and later thawed for continued cultivation. Frozen stock cultures are necessary prerequisites for standardized *in vitro* toxicology testing where frozen stocks of the same passage of a cell line are maintained independent of time and place. Depending on the extent of shared use and availability to other investigators, the procurement of an ultralow temperature freezer is useful. These units maintain temperatures just below the critical point of ice formation (−130°C) and are equipped with alarms and back-up power units in the event of mechanical failure. For independent laboratories, a liquid nitrogen (LN_2) refrigerator is convenient but requires a standing order for the liquid due to constant evaporation.

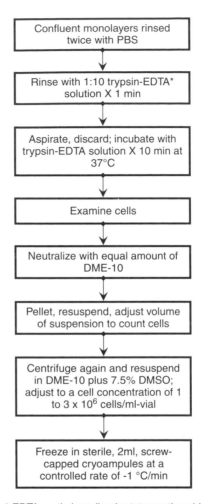

* EDTA = ethylenediaminetetraacetic acid

FIGURE 12.12 Outline of protocol for freezing cell lines. DME-10 = Dulbecco's modified Eagle's medium containing 10% fetal bovine serum. DMSO = dimethylsulfoxide. PBS = phosphate buffered saline. Asterisk = 1:10 dilution of 10X trypsin–EDTA concentrate in PBS.

Cell lines and primary cultures are cryopreserved when the optimal conditions for the culture are attained. The procedure for subculturing of monolayers is launched as described above and outlined in Figure 12.12. Monolayers are dispersed by immersing in a buffered trypsin solution, neutralized with growth medium containing serum (complete growth medium), and centrifuged ($400 \times g$ for 10 min). The supernatant is discarded and the pellet is resuspended in complete growth medium containing an antifreeze such as dimethylsulfoxide (DMSO) or glycerin. Cells are resuspended in DME-10 plus 7.5% DMSO at a concentration of 1 to 3×10^6 cells/ml in sterile 2-ml screw-capped cryoampules. The ampules are labeled and slowly frozen at a controlled rate of $-1°C$/min in an automated freezing chamber. These chambers are usually available as accessory units accompanying liquid nitrogen refrigerators. Alternatively, the ampules are frozen in LN_2 vapors by placing them in a styrofoam container and positioning it in the neck of the LN_2 refrigerator overnight. It is important that the level of the nitrogen is not so high it will freeze the cells too rapidly. The next day, the ampules are snapped into cryocanes and immersed in the LN_2.

Adequate cell stocks of each frozen passage are maintained and a representative vial of cells is thawed and cultured soon after freezing to evaluate the freezing process. This ensures against the loss of a line should contamination or another unanticipated event occur. An ampule is thawed rapidly by agitating in water at 37°C for not more than 3 to 4 min and then swabbed with 70% alcohol, allowed to dry, and the contents are dispensed into warm growth medium using a sterile plugged Pasteur pipette. For most cell types, the culture may be left undisturbed in the incubator in the appropriate atmosphere overnight, after which the medium is replaced.

12.4.6 CELL IDENTIFICATION

In a laboratory that is accustomed to handling several cell lines of many individuals, it is necessary to verify a cell line once initiated, especially when in the presence of risk of cross contamination of cell lines. Some methods used for identification of characteristic cell markers include labeling with fluorescent antibodies, cytogenetic karyotypic studies, isoenzyme profiles, and expression of gene primers using RT-PCR. It is also occasionally necessary to confirm the absence of mycoplasma infections of cultures. This microbial contamination is described as bacterium that lacks a cell wall, tenaciously adheres to cell membranes, and is not eliminated by routine addition of antibiotics to the media.

In fact, the indiscriminate addition of antibiotics to growth and saline-buffered wash media is responsible for routine mycoplasma contamination. The presence of mycoplasma is easily confirmed with the availability of commercial RT-PCR kits containing primers for gene expression of mycoplasma-specific sequences (see websites listed in Suggested Readings at the end of this chapter).

12.4.7 LARGE SCALE BIOPROCESSES

Large scale engineering efforts in mammalian cell culture have made it possible to produce therapeutic biologics including vaccines, ECM components, skin and tissue grafts for surgical manipulations and wound healing, antibodies and hybridomas, diagnostics, and other products for research and clinical applications. These industrial cell culture methods have improved productivity of metabolically engineered cells, increasing the properties favorable for large-scale bioprocessing.

The fundamental physiological basis for cell engineering involves a set of characteristics that select for favorable growth and proliferation and for properties of cellular mechanisms including metabolism, protein processing, and pathways of apoptosis. Some cell culture engineering developments incorporate enhanced process efficiency procedures for large scale cultures of animal cells such as bioreactor processing, feedback culture, and dynamic nutrient feeding. These principles and methods are further described in articles and books listed in the Suggested Readings and Review Articles sections at the end of this chapter.

12.4.8 SOURCES OF CELL LINES

Premier sources of cell lines, cell strains, and other organisms for cell biology and genetics studies are certified cell banks such as the American Type Culture Collection (ATCC, Manassas, Virginia), and the European Collection of Cell Cultures (ECACC, Porton Down, Wiltshire, U.K.). These facilities operate as non-profit institutions and stock thousands of cell types. The ATCC classifies cultures as certified cell lines (CCLs) or cell repository lines (CRLs). More information is available about CCLs: donor, karyotypic analysis, expected population doublings, cell structural features, and enzyme and gene markers. In addition, both cell banks house repositories for viruses, fungi, bacteria, genetic probes, and vectors. Other cell repositories and federal registries are noted below.

SUGGESTED READINGS

American Type Culture Collection (ATCC), Manassas, VA, http://www.atcc.org.
Andrei, G., Three-dimensional culture models for human viral diseases and antiviral drug development, *Antiviral Res.*, 71, 96, 2006.
El-Ali, J., Sorger, P.K., and Jensen, K.F., Cells on chips, *Nature*, 442, 403, 2006.
European Collection of Cell Cultures (ECACC), Porton Down, Wiltshire, U.K. http://www.ecacc.org.uk.
Hay, R.J., Human cells and cell cultures: availability, authentication and future prospects, *Hum. Cell*, 9, 143, 1996.
National Disease Research Interchange (NDRI), Philadelphia, http://www.ndriresource.org.
National Institutes of Health, Human Embryonic Stem Cell Registry, Bethesda, MD, http://stemcells.nih.gov/research/registry.
Seth, G., Hossler, P., Yee, J.C., and Hu, W.S., Engineering cells for cell culture bioprocessing: physiological fundamentals, *Adv. Biochem. Eng. Biotechnol.*, 101, 119, 2006.
Williams, D.J. and Sebastine, I.M., Tissue engineering and regenerative medicine: manufacturing challenges, *IEE Proc. Nanobiotechnol.*, 152, 207, 2005.
Wlaschin, K.F. and Hu, W.S., Feedbatch culture and dynamic nutrient feeding, *Adv. Biochem. Eng. Biotechnol.*, 101, 43, 2006.

REVIEW ARTICLES

Alonso, L. and Fuchs, E., Stem cells of the skin epithelium, *Proc. Natl. Acad. Sci. USA,* 100, 11830, 2003.
Bradley, A., Evans, M., Kaufman, M., and Robertson, E., Formation of germ-line chimaeras from embryo-derived teratocarcinoma cell lines, *Nature,* 309, 255, 1984.
Butler, M., Animal cell cultures: recent achievements and perspectives in the production of biopharmaceuticals, *Appl. Microbiol. Biotechnol.*, 68, 283, 2005.
Evans, M., and Kaufman, M., Establishment in culture of pluripotential cells from mouse embryos, *Nature,* 92, 154, 1981.

Even, M.S., Sandusky, C.B., and Barnard, N.D., Serum-free hybridoma culture: ethical, scientific and safety considerations, *Trends Biotechnol.* 24, 105, 2006.

Findikli, N., Candan, N.Z., and Kahraman, S., Human embryonic stem cell culture: current limitations and novel strategies, *Reprod. Biomed. Online,* 13, 581, 2006.

Garner, J.P., Tissue engineering in surgery, *Surgeon,* 2, 70, 2004.

Kim, J.B., Three-dimensional tissue culture models in cancer biology, *Semin. Cancer Biol.*, 15, 365, 2005.

Kuwahara, M. et al. *In vitro* organogenesis of gut-like structures from mouse embryonic stem cells, *Neurogastroenterol. Motil.*, 16, 14, 2004.

Mackenzie, I.C., Stem cell properties and epithelial malignancies, *Eur J. Cancer,* 42, 1204, 2006.

Rosolowsky, M., McKee, R., Nichols, W., and Garfinkle, B., Chromosomal characterization of MRC-5 cell banks utilizing G-banding technique, *Dev. Biol. Stand.*, 93, 109, 1998.

Thorgeirsson, S.S. and Grisham, J.W., Overview of recent experimental studies on liver stem cells, *Semin. Liver Dis.*, 23, 303, 2003.

Turksen, K. and Troy, T.C., Human embryonic stem cells: isolation, maintenance, and differentiation, *Methods Mol. Biol.*, 331, 1, 2006.

Verlinsky, Y., et al., Repository of human embryonic stem cell lines and development of individual specific lines using stembrid technology, *Reprod. Biomed. Online,* 13, 547, 2006.

13 Cell Culture Methods for Acute Toxicology Testing

13.1 INTRODUCTION

Data obtained from animal experiments yield information pertaining to the dose for lethal or sublethal toxicity that corresponds to many different general toxic mechanisms and effects. Similarly, *in vitro* cell systems also detect a wide spectrum of unspecified mechanisms and effects. In contrast to animal experiments, however, all cell tests of acute toxicity in current use measure the concentration of a substance that interferes with or alters components, structures, or biochemical pathways within cells.

This range of injury is further specified by the length of exposure to the chemical (incubation time), thus allowing a test to be predictive for risks associated with toxic effects *in vivo* for those doses of the tested substance. This assumes that similar concentrations are measured in corresponding human tissue for comparable exposure periods. When undifferentiated (or dedifferentiated) cell lines are used for acute testing, the results are limited to determining the risk for the basal cytotoxicology* of the measured concentrations. Conversely, the use of primary cultures of differentiated cells (such as those of liver, lung, heart, or kidney) yields information capable of predicting cellular toxicity for each organ.

13.1.1 TESTING PRINCIPLES

Cell tests of acute or chronic† toxicity used to mimic local or systemic effects are different from cell tests of mutagenicity and carcinogenicity. The latter have the ability to suggest a known underlying mechanism such as mutation, transformation, initiation, promotion, or progression based on bacterial cell responses to a chemical. They are designated, therefore, as short-term tests and are applied to the prediction of mutagenicity or carcinogenicity of chemicals in humans or animals.

In comparison, assessing the potential carcinogenicity or mutagenicity of a chemical retrospectively in humans normally takes much longer and is plagued with clinical and ethical considerations. Acute cytotoxicology tests instead simulate injuries from tested substances caused by a number of incomplete mechanisms during

---

* The effects of a chemical on fundamental processes that involve structures and functions common to all mammalian cells including membranes, mitochondria, and chromosomes among other organelles.
† The terminology used in describing *in vitro* cell culture models such as acute or chronic has generally replaced the traditional terms of general toxicity.

periods of exposure that are realistic for acute toxicity. This allows for direct extrapolation of the data from the quantitative test to the analogous *in vivo* situation. Thus, although cytotoxicology tests may be acute, they are not short-term. In addition, they are less expensive and time-consuming in comparison to animal tests.

There are similarities between short-term mutagenicity tests and cytotoxicology tests in cell culture. The latter tests are performed using a few standardized techniques whose relevance to human toxicity is pertinent. Also, the concentrations ultimately correspond to human toxic or lethal blood concentrations, whether acute, chronic, systemic, or local. For mutagenicity tests, the relevance applies to the mechanisms that are appropriate for humans.

Historically, cell cultures are employed in mechanistic studies identifying the underlying mechanisms of toxicity or action in response to exposure of cultured cells to low concentrations of chemicals or drugs. Because the development of alternative models for cytotoxicology testing is newer, the protocols are often compared to those of mechanistic studies. As a result some differences between studies of toxic mechanisms in cell cultures and cytotoxicology testing have prompted interesting discussions. In mechanistic studies for example, the parameters and systems vary, depending on the desired information, such that empirical and statistical relevance of the system is secondary to understanding the mechanism of toxicity. With cell methods for acute testing, however, rigorous inter-laboratory evaluation prior to standardization proposals is essential and the procedures have well defined criteria for toxic assessment such as standards for cell viability and cell function. In addition, the protocols are exacting so as to ensure minimal variability.

13.1.2 CRITERIA

Many methods used for assessing systemic and local toxicity are available, some of which are derived from traditional mechanistic models. Their validity was usually established as a result of frequent use in one laboratory, followed by repetition and confirmation by others. This method of evaluation thus proceeds more as a matter of convenience for a laboratory, particularly if the assay is easily accommodated to routine testing facilities. The methods are necessarily similar cells are exposed to different concentrations of a test substance for a specific time period, after which the degree of inhibition of viability or functional parameter is measured. These criteria for standards of measurement represent the toxic endpoint. The most common criteria are listed in Table 13.1.

Some indicators of toxicity detect the net effects of different chemicals and have the advantage of demonstrating a toxic endpoint that is common to many types of mechanisms. Such indicators include damage to cell membranes or loss of viability — generally referred to as a viability [mitochondrial reduction (MTT) or neutral red uptake (NRU)] assay. Other criteria are more selective and sensitive, but are less confident measurements of decreased cell viability, such as measurement of glycolytic pathways or specific enzyme induction or activity. Inhibition of cell proliferation is a sensitive indicator for cellular response to a chemical, especially when coupled with measurements of metabolism.

TABLE 13.1
Basal Cytotoxic Criteria for Assessment of Acute Systemic or Local Toxicity in Cultured Cells

Criteria	Methods Used to Determine Acute Toxicity in Cultured Cells
Cell culture characteristics	Plating efficiency, PDL, loss of phenotypic appearance
Morphology	Phase contrast microscopic analysis
Proliferation	Cell counting, mitotic frequency, DNA synthesis
Metabolism	Glycolysis, protein and lipid synthesis, enzyme activity
Cytoplasmic membrane	Leakage of markers, retention of dyes
Mitochondria	Mitochondrial integrity, cell viability (MTT assay)
Lysosomes	Uptake of dyes (NRU assay)
Chromosomal evaluation	Karyotypic analysis, SCE

Note: MTT = mitochondrial reduction assay. NRU = neutral red uptake. PDL = population doubling level. SCE = sister chromatid exchange. See text for descriptions of assays.

Simultaneous measurements of proliferation and viability are standard indicators of cell integrity, and together with the data from metabolic experiments, contribute significantly to the ability of a cell culture system to predict or screen for toxicity. Cell proliferation, however, is not reliable for demonstrating toxicity in cell lines that divide slowly, such as freshly prepared primary cultures, or cells obtained from adult donors or non-dividing cells. Consequently, the criteria include parameters determined to be fully intact or operational so as to form the basis of a dose response curve.

The most readily detected effects following the exposure of cells to toxicants are phase contrast observations for morphological alteration of the monolayer at low power or observation of individual cells at higher magnifications. Different toxic effects also require investigative tools at higher levels of sensitivity. Gross modifications such as blebbing, vacuolization, and accumulation of lipids are observed using light microscopy and require further histological processing, whereas ultrastructural modifications necessitate analysis by transmission, scanning electron microscopy, or immunohistochemical observation.

Another indicator of toxicity is altered cell proliferation in which the effects of chemicals on the ability of cells to replicate serve as indices of toxicity. The median inhibitory dose (IC_{50}) is the concentration of the test substance at which 50% of the cells do not multiply. Figure 13.1 is a graphic representation of a typical concentration effect curve when proliferating cells are exposed for specific periods. The IC_{50} (0.43 mg/ml), IC_{75} (0.6 mg/ml), IC_{30} (0.3 mg/ml), and IC_{15} (0.15 mg/ml) are calculated from the graph by extrapolation. The three latter concentrations represent cutoff values corresponding to 25%, 70%, and 85% of control.

Another specific measure of replication is plating efficiency, that is the ability of cells to form colonies after 10 to 15 days in culture in the presence of a toxic agent. The information obtained for this parameter, together with cell proliferation, yields more complete information on cell survival, replication, and recovery from toxic

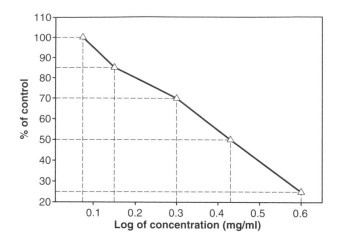

FIGURE 13.1 Typical graded concentration effect curve for a test chemical on cultured cells after 72 hr-exposure. The measured parameter (percent of control) is cell proliferation.

insult. Cell proliferation is measured using several methods including cell counting, DNA content, protein content, and enzyme activity. The assays for DNA content using biochemical methods and monitoring the incorporation of radio-labelled precursors into DNA represent two different methods supplying similar information.

Another general index of toxicity is cell viability. This endpoint is measured by using vital stains such as trypan blue, which enters only compromised membranes of dead cells, and neutral red uptake, which is actively absorbed by living cells. The latter is commonly used for biomaterial testing in the agar overlay method and in cell tests for acute systemic toxicity. Counts of dead and vital cells are compared to controls and provide indices of lethality of the test compound.

Cells derived from different organs or tissues that retain some specialized functions *in vitro* or maintain specialized structures are also widely used in *in vitro* toxicology. For cells that retain differentiated features, effects on more specialized functions and structures are monitored in addition to effects on fundamental processes, including the measurements for specific end products, metabolic pathways, and membrane integrity. Table 13.2 lists some general applications of specific endpoints used for primary and continuous cells that have retained some features from the organ of origin.

Because some cell systems do not possess efficient metabolic activities or have dedifferentiated* with continuous passage *in vitro*, testing in culture may require some form of metabolic activation usually performed in one of three ways:

1. The addition of S9 fractions from rat liver. Mixed function oxidases are induced in cultured cells by treating animals with phenobarbital or β-naphthaflavone before isolation of primary cultures

---

* Mortal or immortal cells maintained in continuous culture through several passages lose some of their original genotypic and phenotypic characteristics while retaining organ-specific functions.

TABLE 13.2
Toxicology Indicators for Cells Retaining Some Differentiated Features in Culture

Descriptive Method	Examples of Indicators Monitored
Synthesis, release, or incorporation of specific molecules	Protein, carbohydrate, and lipid synthesis, metabolic products
Synthesis, release, or activity of specific enzymes	Enzyme release or activity
Interactions of chemicals with cells	Intracellular accumulation, phagocytosis, mitogenic response
Alterations of metabolic pathways	Effect on carbohydrate and lipid energy storage or utilization
Cell surface activities	Receptor binding, membrane polarization, chemotactic response

2. Pre-incubation of the test substance with a primary hepatocyte culture and addition of pre-incubated medium to the test culture
3. Co-culture of target cells with hepatocytes in the presence of the test substance

In addition to the differences in toxic criteria, cell culture methods incorporate different incubation and exposure times and require varying media volumes, cell densities, serum concentrations, and gas phases. Most experiments are performed in 6-, 12-, 24-, 96-, and 384-well microtiter plates (384-well plate has a capacity of 0.1 ml/well). This allows the use of microliter quantities of the test substances and permits easier automation of procedures. The test solution is diluted in serial or logarithmic steps using a microdiluter and many cultures are analyzed simultaneously with plate scanners with spectrophotometric, fluorescent, or luminescent capabilities (see Figure 13.2).

One critical variable during the course of an experiment is exposure time. As with *in vivo* studies, longer exposure times require lower doses of a chemical than the concentrations needed to cause acute toxicity. Toxicity for many substances therefore may increase 10- to 100-fold if the toxic endpoint is measured after 96 hr rather than after 24 hr.

Other factors that alter toxic concentrations include serum component in the medium and whether the cultures are proliferating or static. For instance, high serum concentrations generally decrease toxicity of substances that are protein-bound. Confluent monolayers with established intercellular contacts show more resistance to otherwise comparable toxic concentrations than cells that are in the logarithmic phase of proliferation or in primary cultures.

The physicochemical properties of test compounds influence exposure conditions and the concentrations of toxic agents in the culture medium. The medium is usually an aqueous salt solution containing various soluble additives. Only hydrophilic test compounds are completely solubilized. For volatile toxic agents, special incubation steps are necessary to ensure a constant exposure with time, provided

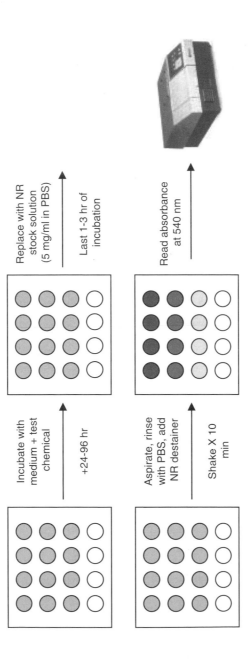

FIGURE 13.2 Outline of NRU assay procedure. Cells are incubated with increasing concentrations of test chemical (gray fill circles) plus blanks (clear fill circles, no cells). NR = neutral red. NR destainer = 1% acetic acid/50% ethanol. PBS = phosphate buffered saline. Photo shows BioTek FLx800 Multi-Detection Microplate Reader. (Courtesy of www.biotek.com.)

the partition coefficients and solubilities are known. Similar problems apply to hydrophobic test compounds and mineral particulates. Lipophilic substances are generally insoluble in water solutions but are solubilized with ethanol, acetone, or dimethylsulfoxide (DMSO) before addition to the medium. Under these conditions, an additional control group receiving the carrier solvent alone is usually included in a study. These practical considerations are further discussed in Chapter 18.

13.1.3 VIABILITY ASSAYS

In order to increase comparability of results and optimize testing procedures, standardization, and adherence to the experimental protocols are desirable. The use of well characterized cell lines, possibly of human origin, is important. Specific toxicity endpoints and the most suitable assay methods are uniformly incorporated into cytotoxicology screens. Cell cultures are examined periodically for possible contamination with microorganisms, for cross contamination with other mammalian cell types, and for genetic integrity.

Adequate documentation provides details of exposure conditions and information on the purity and source of a test compound. Moreover, measurement of the concentration of the compound at the beginning and at the end of the experiment, particularly for insoluble or volatile substances, facilitates inter-laboratory comparisons of results. Two frequently used microtiter methods, the NRU and MTT assays are described below. Both are suitable for rapid screening of proliferation and viability in primary and continuous cell lines.

13.1.3.1 NRU Assay

The NRU cytotoxicology assay is currently in validation studies to determine its suitability for use as an alternative for the prediction of acute systemic toxicity *in vivo* (see ICCVAM in the Review Articles section at the end of this chapter). The assay uses neutral red (3-amino-7-dimethylamino-2-methylphenazine hydrochloride), a weak cationic dye, that is preferentially absorbed into lysosomes. The results rely on the premise that a cytotoxic chemical, regardless of site or mechanism of action, interferes with normal lysosomal uptake, which reflects the number of viable cells. Since only viable cells are capable of maintaining the lysosomal process intact, the degree of inhibition of viability in proportion to the concentration of the test compound provides an indication of toxicity.

Any chemical that demonstrates a selective effect on lysosomes produces a false positive response. Alternatively, this dynamic renders the system useful for detecting chemicals that selectively affect lysosomes and thus more readily suggests a specific mechanism of action. For example, chloroquine phosphate specifically alters lysosomal pH and thus has a greater effect on neutral red uptake than comparably toxic chemicals.

Cells are seeded in microtiter plates and allowed to settle and adhere for 24 hr (Figure 13.2). The growth medium is replaced with fresh incubating medium containing the test chemical at increasing concentrations. Three hours before the end of the exposure period, the medium is aspirated from the wells and replaced with neutral red solution (to a final concentration of 50 µg/ml in the incubating medium). The cultures are incubated for an additional 3 hr at 37°C, after which the medium is removed and the cells are rinsed with PBS, fixed, and destained with acetic acid/50% ethanol. The plates are shaken for 10 min and the absorbance is read at 540 nm with a microplate reader against a reference control and blank wells. Absorbance data correlates linearly with cell viability over a specific optical density range of 0.2 to 1.0.

One major drawback of the assay is the precipitation of the dye into visible, fine, needle-like crystals, yielding inaccurate readings. The precipitation is induced by some chemicals and thus a visual inspection stage during the procedure is important.

13.1.3.2 MTT Assay

In this assay, the tetrazolium salt, 3-(4,5-dimethylthiazol-2-yl)-2,5-diphenyltetrazolium bromide (MTT) is actively absorbed into cells and reduced in a mitochondrial-dependent reaction to yield a formazan product. The product accumulates within cells because it cannot pass through intact cell membranes. Upon addition of DMSO, isopropanol, or other suitable solvent, the sterile product is solubilized, released from intracellular stores, and is readily quantified colorimetrically. Since only viable cells are capable of reducing MTT, the amount of reduced MTT is proportional to the intensity of blueñviolet color formation and greatest cell viability. Thus cell viability correlates with cytotoxicology. This assay is suitable for a variety of culture systems displaying relatively high levels of mitochondrial activity. It should be noted that some compounds selectively affect mitochondria, resulting in n overestimation of toxicity.

The assay begins the same way as the NRU procedure (Figure 13.3). During the final hour of incubation with the test chemical, 10 µl of 0.5% MTT solution is added to the test medium and incubated for 1 to 2 hr at 37°C. The medium plus MTT is then replaced with DMSO and agitated for 5 min. The plates are read at 550 nm with a microplate reader against reference blanks (wells containing no cells). Absorbance correlates linearly with cell number (viability) over a specific optical density range of 0.2 to 1.0. Recently, XTT has been substituted for MTT because XTT is already soluble in aqueous media and obviates the need for DMSO.

13.1.4 CONCLUSIONS

Several multilaboratory validation studies and programs sponsored by regulatory agencies (ICCVAM) have demonstrated that cytotoxicology tests with cell lines yield similar results when the inhibitory concentrations are compared from experiments using similar incubation times and assays. This indicates that different cell types and toxicity criteria measure fundamental processes that allow comparisons of otherwise dissimilar methods.

The relatively small differences among cell lines probably depend on variations encountered with cellular metabolic activity, protocols, and technical variability, leading to differences in the sensitivities of the experimental methods. Thus acute

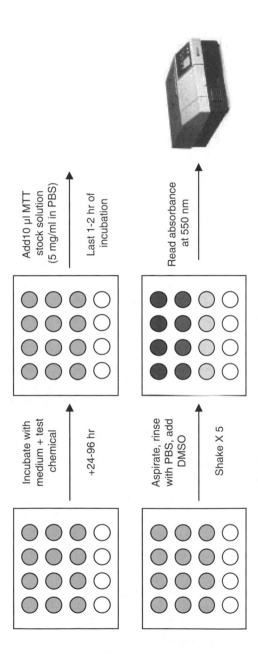

FIGURE 13.3 Outline of MTT assay procedure. Cells are incubated with increasing concentrations of test chemical (gray fill circles) plus blanks (clear fill circles, no cells). MTT = methylthiazol tetrazolium salt. PBS = phosphate buffered saline. Photo shows BioTek FLx800 Multi-Detection Microplate Reader. (Courtesy of www.biotek.com.)

or chronic toxicity detected with these test methods is often referred to as basal cytotoxicology (see Ekwall, B. on Suggested Readings list at the end of this chapter), suggesting that any reasonable method that incorporates an established cell line will accurately measure the fundamental processes inherent in all cells that are affected by toxic insult.

13.2 ACUTE TARGET ORGAN TOXICOLOGY TESTING

13.2.1 USE OF SPECIALIZED PRIMARY CULTURES

Isolation of homogeneous cell populations from the structural framework of an intact organ is a formidable endeavor, yet presents a unique opportunity to study the metabolism of a single cell type. Numerous systems for primary cultures of specialized cell types — cells isolated from an organ as a primary culture that still retain sufficient phenotypic and genotypic features of the parent cells — have been described. Even after several passages, the cultured cells are capable of responding to the toxic effects of compounds as does the *in vivo* target organ of origin. In most cases, the target organs of interest to toxicologists that can explain the mechanistic roles of toxicity include the lungs, liver, skin, kidneys, hematopoietic system, nerve tissue, cells of the immune system, and skeletal and cardiac muscles.

With the exception of techniques that use keratinocytes or ocular models of toxicity, standardized or validated test methods have not been developed to date.* The reason is that primary cultures have several disadvantages for the assessment of organ specific cytotoxicology:

1. Repeated and renewable sources of animals are often needed to establish primary cultures.
2. Isolation of organs and establishing primary cultures are laborious procedures.
3. When used for detecting and quantifying systemic toxicity in target organs, primary cultures are limited to only a few of the many possible organ injuries that could be determined using animals. This is a fundamental limitation of the technology. In conventional whole animal experiments, any or all of the affected organs that show gross pathology are analyzed at the time of sacrifice. The organs are then frozen or preserved indefinitely with fixatives according to the analytical protocol.
4. To standardize testing protocols and generate comparable data, primary cultures are derived from identical animal species and strains.

Development of primary cultures from human donors is also fraught with theoretical, technical, and ethical obstacles. Even as efforts are made to match the ages and sexes of donors, a primary culture established from one individual is not necessarily identical to another. This impediment is overcome by performing the necessary basic

---

* Only select tests for local toxicity are currently in validation programs or have been validated (see Section 13.3).

tests to characterize a primary culture, including karyotypic analysis and identification of functional markers. This battery of procedures in the analytical method ensures progress toward development of standardized methodologies and facilitates inter-laboratory comparisons.

Despite the disadvantages of establishing initial cultures for *in vitro* testing, many types of specialized cells are used to detect target organ toxicities of chemicals. Often, a prototype chemical and its analogues are screened so that their toxicity is compared in the same system. This permits established toxicity data for some of the substances to be used as a reference for previously unknown toxicity of unrelated materials. Hepatocytes fulfill this testing role principally because of the importance that the liver demonstrates in xenobiotic metabolism is the metabolism of exogenous substances. Other cell types that have historical or current utility in target organ toxicology testing include lung macrophages and visceral epithelial cells, blood, cardiac, renal, neuronal, and muscle cells. Procedures describing the isolation of cells from liver, kidneys, and lungs are briefly described below.

13.2.2 ESTABLISHMENT OF PRIMARY CULTURES OF HEPATOCYTES

In vivo, the liver plays an important role in drug and xenobiotic metabolism. The liver performs preliminary screening of most compounds following administration of a drug or chemical. According to the principles of the *first pass phenomenon*, a chemical is susceptible to biochemical metabolic processes immediately after oral absorption because the circulation of the upper intestinal tract is intimately associated with that of the biliary tract. These toxicokinetic processes* alter the chemical structure of the parent compound, rendering it, for the most part, inactive, water soluble, and ready for renal elimination. Thus, the study of toxicokinetics can yield useful information about the disposition of a toxic compound following exposure. The information can be used to predict or monitor the metabolic products of the parent compound, whether toxic or inactive, the duration of the effect, sequestration into body compartments, and the route of elimination.

The process of metabolizing or detoxifying a chemical *in vivo* usually follows a two-step mechanism which, in sequence, results in the formation of a water-soluble, inactive metabolite. The first step in the process, *phase I biotransformation*, involves a group of cellular enzymes including monooxygenases, esterases, and hydrolases, collectively known as the cytochrome P450 enzymes. Through a series of reduction/oxidation reactions, the active sites of chemicals are converted to expose electrophilic polar substituents. With the completion of these reactions, *phase II biotransformation* reactions follow to couple circulating plasma and interstitial constituents to the electrophilic sites of the phase I metabolite. These constituent groups consist of acetyl, glucuronyl, and glutathione moieties that convert the phase I metabolites to water-soluble compounds, thus preparing the non-toxic products for elimination. It is important to note that although the majority of substances are detoxified and

---

* The qualitative and quantitative study of the time course of absorption, distribution, biotransformation, and elimination of an agent in an intact organism.

inactivated through this series of reactions, some substances are activated from non-toxic or inactive parent compounds to active toxic metabolites.*

The fact that the liver has the greatest capacity of all organs to alter chemical substances does not preclude the ability of other organs to contribute to the toxico-kinetics of a compound. The rationale for using hepatocyte cultures, therefore, is based on the abundant metabolic potential of the liver for biotransformation *in vivo*. In addition, the implication is that a drug that exerts an effect on an intact organ will also similarly target the isolated cells. Based on this premise, many of the initial studies using hepatocyte cultures presented conclusions for selected target organ toxicity. Admittedly, target organ toxicity can be correlated with the ability of the liver from a particular species to biotransform an agent and effectively detoxify it. The extrapolation from *in vitro* models to the *in vivo* situation, however, does not depend exclusively on the final products of metabolism nor on the selective role of the liver to alter the chemical, but on the expression of basal cytotoxicology that occurs as a result of accumulation of the compound in the liver. Consequently, hepatocyte cultures are suitable for use as screening methods for systemic toxicity because they can provide information on basal and organ-specific cytotoxicology.

Because of the extensive metabolic capacity of the liver to screen most circulating xenobiotics, hepatocytes have the propensity to screen toxicants that are selectively altered or sequestered by the liver. A significant disadvantage to the use of mortal diploid hepatocytes, however, is that they cannot be routinely maintained as differentiated continuous cell lines and must be established as primary cultures for each experiment. Some of the various methods used for the isolation and maintenance of primary cultures of hepatocytes include *in situ* liver perfusion and digestion, primary hepatocyte enrichment, and hepatocyte cell culture (suspension and attachment cultures).

13.2.2.1 *In Situ* Liver Perfusion

In situ liver perfusion relies on the infusion of fluids through a freshly isolated intact rat liver and collecting the cells in the effluent at the end of the last wash step. Initially, a *wash perfusion* medium formulated in HBSS and containing heparin as the active ingredient is infused through the through the inferior vena cava at 25 to 35 ml/min through one arm of a Y tube connector. The solution removes blood from the portal circulation. After 10 min, the *digestion perfusion* medium, also formulated with HBSS containing collagenase, is infused through the other arm of the connector for 10 min. The liver is then transferred to a petri dish and 20 ml of a fresh cell wash medium is injected into the organ.

Further release of hepatocytes is stimulated by mechanically agitating the dish, surgically manipulating small incisions in the organ, and scraping the lobes of the liver. The procedure is repeated and the solutions are collected and centrifuged. Primary cultures are initiated by inoculating culture flasks with the cell suspensions. All solutions are prepared, and manipulations are performed under aseptic conditions.

* In fact, the mechanisms of many therapeutic drugs, especially cancer chemotherapeutic agents, rely on the conversions of inactive parent compounds to active metabolites.

Perfusion of explanted liver specimens from a human donor requires performance of aseptic technique in a manner not unlike that described for *in situ* perfusion. Briefly, fresh biopsy material is perfused through ostensibly exposed blood vessels on the outer surfaces of the organ. Perfusion, digestion, and wash-out media are similar in composition as above, and flow rates are maintained constant and recirculated. Further release of primary hepatocytes is accomplished by gentle combing and shredding of the intact organ after transferring to sterile petri dishes.

13.2.2.2 Primary Hepatocyte Enrichment

Suspensions of isolated liver cells are manipulated so as to remove non-hepatocyte and non-viable cells from the desired cultures. A low-speed, isodensity Percoll™ centrifugation procedure used to enrich the population of primary hepatocytes from liver cell isolates has been described (Gomez-Lechon, M.J. et al.). Using aseptic technique, a suspension of freshly isolated and washed cells is mixed with and then layered on a stock solution of iso-osmotic Percoll and centrifuged at $50 \times g$ for 10 min at 4°C. The parenchymal hepatocytes are separated and removed from the bottoms of the tubes while contaminating and non-viable cells float on top. The primary cells are then resuspended in appropriate medium in preparation for primary cultures.

13.2.2.3 Hepatocyte Cell Culture

As with most cell cultures, freshly isolated hepatocytes are maintained using routine culture procedures. The criteria for selecting any system is based on the degree of differentiation that the primary cultures undergo once exposed to the artificial culture conditions and the consequent loss of differentiated functions. Primary cultures of parenchymal cells suspension cultures are not desirable because these cells result in a rapid loss of differentiated functions and cytochrome P450 levels. Thus cytotoxicology tests must be performed generally within 12 hr.

Attachment cultures, using flasks or wells to which basement membrane substrata have been layered, are more suitable. Establishing primary cultures on plastic cell culture plates coated on cellular attachment components maintains differentiated features, improves viability and membrane integrity, especially after the traumatic isolation procedures, and also preserves cytochrome P450 activity for several weeks. The substrata include collagen types I and IV, laminin, fibronectin, or combinations of these, and are applied as soluble solutions to the culture plate surfaces.

13.2.2.4 Functional Markers of Primary Hepatocyte Cultures

Table 13.3 summarizes some of the important criteria used for assessing functional and differentiated features of freshly isolated liver parenchymal cells. These criteria are also incorporated in cytotoxicology testing of chemicals that exhibit organ-specific toxicities. In fact, one or more test methods are used to screen chemicals for potential liver toxicity. In addition, the ability of a chemical to undergo biotransformation is efficiently evaluated using primary cultures of hepatocytes.

TABLE 13.3
Functional Markers of Primary Hepatocyte Cultures

Functional Marker	Descriptive Indicators
Cell membrane integrity	Trypan blue exclusion, lactate dehydrogenase leakage, aspartate aminotransferase leakage
Phase I metabolic reactions	7-ethoxycoumarin-O-deethylation activity
Phase II conjugation reactions	Glutathione determination, glutathione-S-transferase activity
Peroxisomal-oxidation capacity	Cyanide-insensitive acyl-CoA oxidase activity, carnitine acyltransferase activity

---

13.2.3 ESTABLISHMENT OF PRIMARY CULTURES FROM OTHER ORGANS

Principles of isolation of cells and for establishing primary cultures are described below. Detailed methods for isolation procedures are described in several review articles listed at the end of this chapter.

13.2.3.1 Isolation of Renal Cortical Cells

Several methods effectively yield sufficient quantities and purity of cortical renal cells from a variety of animal species. The cortical cells stem from the renal cortex and constitute the most metabolically active cells of the kidney. Of these, proximal tubule and distal tubule cells are more frequently used for the assessment of *in vitro* renal toxicity. The methods of isolating and obtaining enriched populations of tubular cells rely on the ability of the procedures to selectively exclude other contaminating cell types such as cells associated with the glomeruli and cortical collecting duct. The most popular protocols include enzymatic methods, mechanical methods and historically, microdissection techniques. Any of these may be coupled with density gradient centrifugation on Percoll to enrich the fractions for proximal or distal tubule cells.

Although developed as one of the original methods for isolating cortical cells, *microdissection* has limited use because of poor yields of tubule cells and the prolonged times required. Essentially, the method requires precise manipulative skills using microdissecting capability in order to focus on nephron tubules. The method is protracted and, for the purposes of screening nephrotoxic chemicals, does not afford enough cells for a suitable cytotoxicology testing procedure.

Mild enzymatic digestion with purified collagenase in a balanced salt medium is the method of choice for releasing cortical cells from the basement membrane. In general, the procedure is similar to that described above for enzymatic digestion for hepatocyte isolation.

Mechanical separation of proximal cells relies on the *in situ* perfusion of the kidney with a suspension of iron oxide particles. These micronized pellets lodge in the glomeruli and allow for the separation of proximal tubule cells from the filtering apparatus. Ideally, after perfusion, portions of the cortex are hand homogenized, coarsely filtered, and the suspension is exposed to a mechanical stirring bar to which the iron-laden glomeruli are attracted and separated.

Thus, selection of a particular isolation procedure must primarily yield a sufficient number of cells adequate for toxicity testing and, in addition, the cells must be of sufficient purity so that reliable data are generated. Underlying this premise is the need to assess the functional characteristics of the cell types.

13.2.3.2 Use of Continuous Renal Cell Lines

Over the past 30 years, improved methods for maintaining continuous cultures of cells have allowed for the realization of general mechanistic cytotoxicology studies. Renal cell culture technology has benefited primarily from the improved methodology in much the same way as in other fields. The problems associated with other continuous cell lines, however, have also plagued the maintenance of continuous cultures of renal cells (and include those discussed in Chapter 12, Cell Culture Methodology), the most important of which include the loss of *in vivo* characteristics following isolation and subsequent passage over several generations *in vitro*. This dedifferentiation may be followed by a decrease in the levels of functional markers necessary for identifying the cell type and for assessing cytotoxicology. Nevertheless, propagation of continuous cell lines has allowed for important contributions to understanding renal cell biology and physiology. It may be necessary to approach renal cytotoxicology, however, more cautiously when using cell lines. As with the use of continuous cell lines from any organ, the apparent toxicity from a chemical may be a reflection of basal cytotoxicology to the cell rather than a specific renal toxic effect.

Among the most frequently used permanent cell lines used in renal studies are the transformed cells including MDCK (Madin-Darby canine kidney) and LLC-PK renal epithelial cell lines.

MDCK cells maintain morphological features and functional characteristics consistent with cells of distal tubule/collective duct origin. In addition, they demonstrate biochemical evidence of transporting epithelia, although several strains and clones are available that differ considerably from the parent cell line.

LLC-PK cells are derived from pig proximal tubule cells and share morphological features characteristic of their parent cells, including the ability to form domes, the presence of apical membrane microvilli, and several junctional complexes. Physiologically they express high activities of brush border membrane marker enzymes, retain the ability to transport small molecules, and express high transepithelial electrical resistance (TEER) when cultured on microporous membrane filter culture inserts.

Assessment of functional features for identification of primary and continuous cultures of differentiated renal cortical cells include general cell culture markers for membrane integrity and cell proliferation and morphological examination for phenotypic appearance. Assessing the abilities of cells for toxicokinetic studies requires intact and sufficiently high levels of enzymatic activity including the presence of cytochrome P450 isozymes and other oxidative enzymes present in proximal tubular brush borders (phosphorylases, dehydrogenases, ATPases). Collectively, these features provide evidence for the origins of the cells and lend support for data generated during metabolic investigations.

13.2.3.3 Establishment of Primary Cultures of Lung Cells

Mechanistic studies of organ-specific toxicity in mammalian lungs have been hampered by the heterogeneity of lung composition. Over 40 different cell types are identified in this organ, not all of which have been extensively characterized. This complex cellular organization, together with cell interactions with the extensive vascular, lymphatic, and neuronal networks, presents formidable challenges to the performance and interpretation of mechanistic studies of lung-specific toxicity. Table 13.4 lists some of the major cell types currently used in pulmonary studies, along with their unique biochemical features. These cells have also found utility in *in vitro* cytotoxicology studies.

TABLE 13.4
Cell Types of Mammalian Lungs Isolated in Cell Culture

Cell Type	Origin	Structural *In Vitro* Features	Functional Characteristics
Alveolar type II pneumocyte	Alveolar epithelium	Irregular cuboidal cells, show microvilli and lamellar bodies	Synthesis and storage of pulmonary surfactant; xenobiotic metabolism
Alveolar type I pneumocyte	Alveolar epithelium	Squamous epithelial cells, long cytoplasmic extensions	Gas exchange
Clara cell	Bronchioles	Non-ciliated bronchiolar epithelial cell with electron dense secretory granules	Secretory, xenobiotic metabolism
Ciliated bronchiolar epithelial cell	Bronchiolar, tracheal origin	Ciliated epithelial cell with electron dense secretory granules	Regeneration and repair
Endothelial cells	Vascular origin	Cuboidal appearance with caveolae* on luminal surface	Gas exchange, solute transport, secretory, immune cell migration
Alveolar macrophage	Alveolar interstitium and epithelial surface	Large, mononuclear phagocytes, numerous cytoplasmic extensions	Phagocytic, migratory, immune activation

* Sarcolemmal vesicles usually associated with sparse sarcoplasmic reticulum in smooth muscle cells; function in release and sequestering of calcium.

The isolation and maintenance of relatively pure populations of alveolar and bronchiolar epithelial and endothelial cell lines from mammalian lungs utilize enzymatic digestion of normal rodent lung followed by centrifugal elutriation, laser flow cytometry, or differential adhesion procedures. Type II pneumocytes have been maintained in continuous culture following isolation from rats, rabbits, cats, hamsters, and mice. They retain high levels of enzymatic activity, display cytoplasmic lamellar inclusions, and synthesize and secrete surfactant. These features are maintained in culture for many weeks and, as a result, are used in mechanistic cytotoxicology studies for pulmonary alveolar specific toxins. Derivation of continuous cell lines of alveolar epithelial cells from primary cultures, however, has generally resulted in the disappearance of characteristic features such as the loss of cytochrome P450 enzyme activity and reduction in proliferation within 2 wk of initial isolation.

The *organotypic culture* represents a model system suitable for the investigation of *in vitro* cytotoxic responses. Isolated cells are layered on Gelfoam squares, reaggregate, and form alveolar-like structures, thus resembling their *in vivo* appearance. Generally the cultures are established from fetal, newborn, and postnatal rats, retain differentiated characteristics for longer periods of time in culture, and maintain their capacities to produce and secrete pulmonary surfactant. Organotypic models of type II pneumocytes are useful for identifying agents with potential capacity for oxidant lung injury.

Clara cells are examples of non-ciliated bronchiolar epithelial cells that possess a secretory role. Ultrastructurally, as with pulmonary endothelial cells, the presence of extensive endoplasmic reticulum, abundant mitochondria, and numerous opaque osmiophilic* granules suggests a high level of metabolic activity. The fact that metabolically active cells are targets for chemical injury is related to the presence of xenobiotic metabolizing enzymes. Accordingly, both Clara cells and type II pneumocytes have been shown to play a major role in detoxification of xenobiotics as a result of their enzymatic capacities.

Macrophages constitute one of the major classes of mononuclear phagocytes originating from bone marrow. Alveolar macrophages migrate, circulating principally in the capillaries of the lungs and spleen. The cells are approximately 20 μm in diameter and are distinguished by their unlobed nuclei, microvilli, and absence of neutrophilic granules. Macrophages play an important role in immune response and resistance to infection. They are phagocytic, engulfing foreign particles and cellular debris into intracytoplasmic vacuoles. They are also instrumental in activation of lymphocytes in humoral and cell-mediated immunity.

Alveolar macrophages are obtained from rodents by pulmonary lavage, through repeated instillation and aspiration of sterile balanced salt solution in the trachea. Cells are isolated from collected lung fluids after centrifugation. Cultures are identified and characterized morphologically and cytochemically for the demonstration of macrophage-specific esterases.

The pulmonary alveolar capillary network lies in the submucosa beneath the basement membrane and comprises about 20% of the total body capillary system. These vessels are lined with fenestrated and non-fenestrated endothelial cells that

---

* They appear as dense cytoplasmic opaque granules when reacted with osmium tetroxide.

function in gas exchange and movement of solutes. The cells regulate the transport into the circulation of important chemical mediators such as biogenic amines, kinins, angiotensins, and prostaglandins. In addition, they monitor migration of monocytes, granulocytes, neutrophils and lymphocytes, from the lymphatic to systemic circulations.

Endothelial cells are ideally suited to conduct a variety of specific metabolic activities important in the pulmonary processing of vasoactive substances. Enzymes, chemically mediated inhibitors, cytoplasmic receptors, and transport systems of endothelial cells determine the level of hormones entering the circulation. Through the regulation of these mechanisms the pulmonary endothelium plays an important role in the response of the lung to xenobiotics.

Endothelial cells from large and small vessels of the lung possess specialized properties and characteristics. The presence of fenestrations, organ-specific antigens on surface membranes, and cell membrane-specific glycoproteins are examples of differences among cells of various origins. The cells are generally isolated by enzymatic digestion of the greater vessels from rodent or rabbit lungs. Cells grow to form confluent monolayers using routine cell culture techniques.

In culture, endothelial cells demonstrate typical cuboidal arrangements when confluent and they possess caveolae on their luminal surfaces. Functionally, detectable activity of angiotensin converting enzyme (ACE) on the plasma membrane and intracytoplasmic xenobiotic metabolizing enzymes are hallmarks of the cells in culture.

13.2.4 DIFFERENTIAL TOXICOLOGY TESTING

It is important to note that differentiated cells derived from passages of primary cultures are inclined to reflect both organ-specific and indirect effects of basal cytotoxicology when challenged with a toxic insult. It is possible, however, that by studying specialized functions of organ-specific cells, fine adjustments in the dosages used during exposure are required to distinguish basal cytotoxicology from target organ effects. Alternatively, toxicity is demonstrated by selecting a narrower range of dosages or incubation for shorter periods of exposure. This requires analysis at intervals *during* the exposure period rather than at the end of the incubation.

13.3 ACUTE LOCAL TOXICOLOGY TESTING

No other field in *in vitro* toxicology testing has driven academic, industrial, and government resources to develop cell modeling systems as much as the need for alternatives to local toxicity testing. This is principally due to the ubiquitous presence of cosmetics, toiletries, and dermal and ocular pharmaceuticals marketed for worldwide use. Furthermore, the results obtained with the original models for dermal and ocular toxicity were not successful in predicting effects encountered *in vivo*, primarily because of the inconsistencies between data generated from animal models and human risk exposure. Recent technical advances and improved knowledge of cell culture technology have allowed improvements in the development of cellular systems, with some tests undergoing full validation evaluation in the United States (US) and European Union (EU).

The development of a cellular model of local toxicity involves the establishment of primary cultures from specialized tissues, especially dermal and ocular epithelium. These primary cultures possess specific functional markers that can distinguish selective toxic effects. In particular, the effects of a chemical on a specific functional target that is unique for ocular epithelia is assessed and distinguishes the chemical's effect on ocular cells from effects on other cell types. Other isolated *in vitro* models used to screen for local toxicity incorporate tissue types that are not limited to organ of origins and include supportive structures such as connective tissue cells. For instance, corneal epithelial cell culture models have been introduced to test for chemicals with potential for detecting ocular effects.

Most *in vitro* tests in development involve cell and tissue cultures of human or animal origin. The systems are used to evaluate the effects of substances in isolated environments devoid of hormonal, immune, or neural influence. These advantages have also been noted as drawbacks — the elimination of other biological factors in the system does not allow the method to mimic interactions occurring in the whole organism. Nevertheless, many *in vitro* methods have been developed since the early 1990s; widespread regulatory acceptance, however, especially in the US, is limited. The problem centers on the time and cost necessary to develop the tests and the arduous validation process.

13.3.1 OCULAR TOXICOLOGY TESTING

The current *in vivo* Draize rabbit eye test method identifies both irreversible (e.g., corrosion) and reversible ocular effects. It also provides scoring that allows relative categorization of severity for reversible effects such as those caused by mild, moderate, or severe irritants. The Draize test, however, has been criticized for its lack of reproducibility, its subjective nature of assessment, the variable interpretation of results, the high dose of test material used, the over-prediction of human response, and the infliction of pain to animals. The normal endpoints of toxicity for the *in vivo* rabbit eye test include corneal opacity, inflammation, and cytotoxicology.

Current EPA ocular testing guidelines and the United Nations (UN) Globally Harmonized System (GHS) of Classification and Labeling of Chemicals (2003) indicate that if serious ocular damage is anticipated,* then a test on a single animal may be considered. When anticipated damage is observed, no further animal testing is necessary (EPA, 1998; UN, 2003); if it is not observed, additional test animals may be evaluated sequentially until concordant irritant or non-irritant responses are observed.

One of the original models developed in the early 1990s as a biochemical procedure to evaluate ocular irritation was the EYTEX™ system. The test was based on the concept that the normal transparent state of the cornea depends on the relative degree of hydration and organization of corneal proteins. Corneal opacification,

---

* Serious ocular damage to the rabbit eye is described as irreversible adverse effect still occurring on day 21 after exposure.

TABLE 13.5
Current Models of Alternative Methods for Ocular Toxicology Testing

Name	Validation Status	Test Method Indicator	Test Objective
HET-CAM	Pre-evaluation (US)	Hemorrhage, coagulation, cytolysis	Ocular sensitivity and corrosion
BCOP	Pre-evaluation (US)	Increase in corneal opacity and thickness	Ocular sensitivity and corrosion
ICE	Pre-evaluation (US)	Increase in corneal opacity and thickness	Ocular sensitivity and corrosion
IRE	Pre-evaluation (US)	Increase in corneal opacity and thickness	Ocular sensitivity and corrosion
EpiOcular	Not validated	MTT cell viabiity	Ocular sensitivity and corrosion

Note: BCOP = bovine corneal opacity and permeability. HET-CAM = hen's egg test–chorioallantoic membrane. ICE = isolated chicken eye. IRE = isolated rabbit eye.

therefore, is a result of a decrease in protein hydration associated with changes in protein conformation and aggregation. The EYTEX test simulated corneal opacification by using alterations in the hydration and conformation of an ordered macromolecular matrix to predict *in vivo* ocular irritancy. Although the model did not prove reliable as an *in vitro* prescreening system for determining eye irritation potential of cosmetic formulations, it set the foundation for further development of ocular toxicology testing models.

By 2003, the EPA nominated four *in vitro* ocular toxicity methods for evaluation by the Interagency Coordinating Committee on the Validation of Alternative Methods (ICCVAM, NIEHS):

1. Hen's egg test–chorioallantoic membrane (HET-CAM) assay
2. Bovine corneal opacity and permeability (BCOP) assay
3. Isolated chicken eye test method (ICE) or chicken enucleated eye test (CEET) method
4. Isolated rabbit eye (IRE) assay

ICCVAM endorsed the pre-evaluation of these methods by the National Toxicology Program (NTP) Interagency Center for the Evaluation of Alternative Toxicological Methods (NICEATM) and recommended background review documents (BRDs) for each method. The procedures are used as part of a preclinical battery of tests by many EU countries for identification of ocular irritants. The principles of the procedures and their validation statuses are listed in Table 13.5 and reviewed below.

13.3.1.1 HET-CAM Test

Initial results in the 1990s using a modification of the HET-CAM test, showed a high rank order correlation between the scores from testing chemicals and cosmetic formulations with Draize *in vivo* data. Later studies included classification of several classes of irritants ranging from non-irritant to severe, with high correlations between the *in vitro* model and the Draize test. This breakthrough projected that *in vitro* models could represent valid alternative considerations to predict *in vivo* ocular toxicity.

The procedure monitors blood vessel changes in the chorioallantoic membranes (CAMs) of chicken eggs. The CAM is a vascular fetal membrane composed of the fused chorion and allantois. Briefly, fertilized hen's eggs are incubated under optimized conditions for 9 days. On the 10th day, the eggs are opened and the CAM exposed. The test substance is applied to the surface of the CAM for 20 sec, rinsed with water, and evaluated for development of irritant endpoints (hyperemia, hemorrhage, and coagulation) at 0.5, 2, and 5 min after rinsing. Irritant effects in the CAM blood vessels and albumen are subjectively assessed and a score is assigned based on the time required for development of each endpoint. The scores are totaled to yield a total irritation score (maximum of 21) for the tested substance.

The method is proposed to provide information on effects that may occur in the conjunctiva following exposure to a test substance as a response to local injury rather than monitoring direct cytotoxicology. The test also assumes that acute effects induced by a test substance on the small blood vessels and proteins of this soft tissue membrane are similar to effects induced by the same test substance in the eyes of treated rabbits and correlates to irritation and/or corrosion in human eyes.

The HET-CAM method is currently used by select toxicology testing companies for the identification of ocular corrosives and severe irritants in a tiered testing strategy. Positive *in vitro* test results are considered in a weight-of-evidence decision as to whether to classify the substance as an ocular corrosive or severe irritant. The ability of the test to correctly identify ocular corrosives and severe irritants, as defined by the EPA (1996), EU (2001), and GHS (UN, 2003) is currently in validation studies (ICCVAM, 2006).

In preliminary evaluations, the HET-CAM has shown potential to refine and reduce animal use in eye irritation testing when used in the GHS tiered testing scheme. In addition, when used to assess severe eye damage, this method will reduce the numbers of animals subjected to testing and reduce their pain and suffering by their exclusion from testing of corrosives and severe irritants.

13.3.1.2 Bovine Corneal Opacity and Permeability (BCOP) Assay

The BCOP assay is an organotypic (i.e., isolated whole organ) *in vitro* eye irritation test method developed as a modification of an earlier ocular irritation assay using isolated bovine eyes. The assay provides for short-term maintenance of normal physiological and biochemical functions of the cornea in an isolated system. The basis of the test relies on the role of the cornea as one of the main targets during accidental eye exposures since damage to the cornea can result in visual impairment or loss. In addition, corneal effects are weighted heavily in the *in vivo* Draize eye test scoring system for ocular irritancy (e.g., 80 of 110 points).

The BCOP assay measures opacity and permeability, two important components of ocular irritation. Opacity is determined by the amount of light transmission through the cornea; permeability is measured by the amount of sodium fluorescein dye that passes through all corneal cell layers. Since ocular irritation also involves other tissues such as the iris and conjunctiva, these *in vitro* toxicity measurements represent only one aspect of the overall complex response of the eye to irritants. Recently, several additional endpoints to the BCOP assay have been added as means of assessing ocular toxicity including corneal swelling or hydration and allowing histological evaluation of morphological alterations to determine the type and depth of corneal injury and reversibility of tissue damage.

During the preparation of the model, the cornea is dissected from the eye structure and mounted in a plastic holder fitted into a two-compartment cell capable of measuring light transmission. With the epithelial side of the cornea on the top, the test substance is introduced in the upper chamber, after which transmission is measured spectrophotometrically. Diffusion of fluorescein dye to the lower chamber is also measured and correlated with the opacity index and *in vivo* Draize test scores.

Although the BCOP is not yet validated, the EU national regulatory authorities accept positive outcomes from this eye irritation test method for classifying and labeling severe eye irritants (R41). If a negative result is obtained, an *in vivo* test is subsequently required since the BCOP has not been shown to adequately discriminate between eye irritants and non-irritants (EU, 2004). In addition, the assay is currently used by certain US and EU companies as an in-house method to assess the ocular irritation potentials of a wide range of substances or products. As an *in vitro* assay for ocular toxicology testing, the BCOP method could potentially be used to identify the irreversible, corrosive, and severe irritation potentials of drugs and chemicals in a tiered testing strategy (GHS, UN, 2003).

13.3.1.3 Isolated Chicken Eye (ICE) Test

Based on the IRE test (see below), the ICE test protocol is an organotypic *in vitro* bioassay. A test substance is applied to the corneas of eyes isolated from chickens that have been processed for human consumption. The test substances is applied as a single dose for 10 sec, followed by rinsing with isotonic saline. Following exposure to a chemical substance, the extent of damage to the eye is determined by measuring corneal swelling, corneal opacity, and fluorescein retention. Only analysis of corneal swelling provides a *quantitative* measurement, thus potentially providing improved precision and reduced inter-laboratory variability compared to the *in vivo* rabbit eye test.

As with the BCOP assay, corneal swelling in the ICE test is determined by calculating the increase in corneal thickness from a baseline measurement. Corneal thickness has been identified as a quantitative and reliable endpoint for the evaluation of corneal injury. Fluorescein retention provides an assessment of permeability, indicative of damage to the corneal surface. Thus corneal opacity in both the BCOP and ICE assays provides a measurement of corneal damage that can be directly correlated to the *in vivo* rabbit eye test. Morphological changes and histopathology are also included in the study protocol for a more accurate assessment of extent of corneal injury. Based on the maximum mean values of the measurements, the irritation potential of a test substance is defined within a range from non-irritating to severely irritating. Since the method has not been shown to adequately discriminate between eye irritants and non-irritants, substances test negatively in ICE for severe or irreversible effects undergo additional testing to confirm the possibility of false negatives.

As with the BCOP, the ICE method is not yet validated. EU regulatory authorities, however, accept positive outcomes for eye irritation from this test method for classifying and labeling severe eye irritants (R41). Consequently, the method may be useful in a battery of *in vitro* eye irritation test methods that collectively predict eye irritation potentials of substances *in vivo*.

13.3.1.4 Isolated Rabbit Eye (IRE) Test

The IRE test was developed as an *in vitro* alternative to the *in vivo* Draize rabbit eye test for the assessment of eye irritation. In the IRE method, candidate substances are applied over the corneas of enucleated rabbit eyes in a manner similar to the ICE test.

This method enjoys several advantages that allow it to be considered for validation programs: eyes are selected from animals that are euthanized prior to ocular irritancy testing, from animals used for other toxicological purposes, or from animals in the food chain. Since testing is performed on the cornea, there is good relevance with the Draize test for scoring ocular irritancy. As with the BCOP, the effects of the test substance on the cornea of the isolated eye are measured quantitatively as an increase in thickness (swelling), subjectively as scores for corneal opacity and fluorescein penetration, and descriptively as morphological changes to the corneal epithelium.

The lack of a widely accepted standardized IRE test method for detecting ocular corrosives and severe irritants has hampered its evaluation as a partial or full replacement for the Draize rabbit eye test. Consequently, the method has been modified for use in the assessment of selective types of irritants (e.g., severe irritants) and specific classes of chemical substances (e.g., surfactant-based chemicals and cosmetic and hair care products).

13.3.1.5 EpiOcular™ Model

The EpiOcular (OCL-200, MatTek Corporation, Ashland, Massachusetts) is an *in vitro* human corneal model designed to replace the traditional animal Draize eye test. The model consists of a three-dimensional *in vitro* human corneal epithelium composed of normal human-derived epidermal keratinocytes cultured on a permeable polycarbonate membrane that forms a stratified squamous multilayered epithelium similar to that of the cornea. The tissue construct has an air–liquid interface and exhibits morphological and growth characteristics that mimic *in vivo* conditions.

Cellular viability is measured with the MTT colorimetric assay using an IC50 endpoint for comparison among different potential toxicants. The model provides data measuring cellular cytotoxicology as a reflection of ocular irritation. The test is purported to yield eye irritation data comparable to responses to the *in vivo* rabbit eye test. Although not validated, the test can distinguish among the irritation potential of mildly toxic products.

TABLE 13.6
Current Models of Alternative Methods for Dermal
Toxicology Testing

Name	Validation Status	Test Method Indicator	Test Objective
LLNA	ICCVAM/US	Allergic skin reaction	Skin sensitization
Corrositex	ICCVAM/US	Visual color changes	Skin corrosion
EpiDerm	EU/US	MTT	Skin corrosion
EpiSkin	ECVAM/EU	MTT	Skin corrosion
TER	ECVAM/EU	Electrical resistance	Skin corrosion
3T3 Phototoxicity	ECVAM/EU	NRU	Phototoxicity

13.3.2 DERMAL TOXICOLOGY TESTING

Screening techniques to detect local toxicity are used to investigate the effects of various irritating substances on the skin and in the lungs. Substances studied in lung cell cultures include tobacco smoke, diesel exhausts, nitrogen dioxide, aldehydes, phorbol esters, and paraquat. The isolation of alveolar epithelial cells and non-ciliated bronchiolar epithelial (Clara) cells and their establishment in primary culture greatly accelerated the underlying mechanisms of the local pulmonary responses to xenobiotics. The culturing conditions, however, are not without problems and are often labor-intensive (see previous sections on establishing primary cell cultures). In addition, continuous cultures of these cells are difficult to maintain in their differentiated state. Despite this, the need for isolating dermal and lung cells is especially warranted, particularly because they are known to be instrumental in the inflammatory response.

Earlier attempts to assess skin irritation of chemicals *in vitro* have met with considerable setbacks. The deficiencies associated with dermal fibroblasts as models for skin toxicity studies center around the continuing controversy of the dedifferentiation of the cells once they are maintained in serial culture. They lose their phenotypic appearance and characteristic features of their original parent cells. Recent progress in dermal toxicological studies has concentrated on the development of organotypic models that mimic mammalian skin. Table 13.6 lists current models for dermal toxicology testing.

13.3.2.1 Murine Local Lymph Node Assay (LLNA)

LLNA is a scientifically accepted protocol used to assess the potential of substances to produce skin sensitization. LLNA uses mice as replacements for guinea pigs in the conventional methodology and is based on the visual evaluation of the ability of a chemical to cause allergic reactions after repeated application to the skins of 20 to 40 animals. Although not a true *in vitro* method, the validity of the assay was endorsed by independent peer review evaluation based on its ability to cause less pain and distress to animals (ICCVAM, 2005).

13.3.2.2   Corrositex™ Assay

The Corrositex assay (In Vitro International, Irvine, California) is a standardized and quantitative *in vitro* test for assessing skin corrosion. It is based on the time required for a test chemical to penetrate a biobarrier membrane. The membrane is composed of a reconstructed collagen matrix, developed to mimic the physicochemical properties of rat skin. The time needed for visual color change is inversely proportional to the corrosivity level of the test chemical. Corrositex was granted regulatory approval by the U.S. Department of Transportation (DOT), EPA, Food and Drug Administration (FDA), Occupational Safety and Health Administration (OSHA), and the Consumer Product Safety Commission (CPSC). The method is a replacement for the *in vivo* Draize rabbit skin test and its use is limited to assessing corrosivity of specific classes of chemicals including organic bases and inorganic acids.

13.3.2.3   EpiDerm™, Episkin™, and TER

The European Commission on the Validation of Alternative Methods (ECVAM) has validated three alternative *in vitro* test models for detecting skin corrosivity: EpiDerm (EPI-200, MatTek Corporation, Ashland, Massachusetts), Episkin (Imedex, Inc., Alpharetta, Georgia), and the rat skin transcutaneous electrical resistance (TER) method.

The EPI-200 skin corrosion test is a validated alternative method to the Draize rabbit skin corrosion test in both the US and EU. Episkin is validated for identifying corrosive properties of chemicals. It is an accepted test in the EU and for two of the three UN packing group classifications (groups I and II/III, ICCVAM, 2002).* Both EpiDerm and Episkin use a three-dimensional system consisting of a reconstructed human epidermis and functional stratum corneum. Analysis of the stratum corneum using transmission electron microscopy reveals intercellular lamellae characteristic of normal human epidermis. Unlike the previous models, EpiDerm appears to have hemidesmosomes arranged in a well organized basement membrane. The tests are validated in Canada and the EU as total replacements for animal-based skin corrosion studies. In addition, EpiDerm is the only commercially available organotypic *in vitro* model for assessing dermal corrosivity approved by both EU and US regulatory agencies.

The TER method monitors changes in transcutaneous electrical resistance in a rat skin disk as a result of chemical exposure over time. Substances with resistance values greater than 5 kΩ/cm are considered non-corrosive. The TER method is validated and is part of a battery of tests for skin corrosion in the EU.

ECVAM, in collaboration with the EC and European Cosmetics, Toiletry and Perfumery Association (COLIPA), has completed validation studies on other tests including the 3T3 neutral red uptake (NRU) phototoxicity test and its application to ultraviolet filters (ECVAM, 1998–2001). The *in vitro* NRU test, a standard cell viability assay, uses BALB/c 3T3 mouse fibroblasts to determine the cytotoxic effect of

---

* Group assignment is based on a chemical's ability to cause skin corrosion after 3 min, 1hr, and 4 hr, respectively.

a chemical in the presence or absence of non-cytotoxic ultraviolet radiation. The NRU test is validated in the EU as a replacement for acute animal phototoxicity studies.

13.4 SUMMARY

The development of *in vitro* methods for assessing acute systemic and local toxicity has progressed considerably in the past 20 years as a result of similarities of test methods to the traditional *in vivo* protocols. In fact, the cell culture models for local toxicology testing, in particular, incorporate several features consistent with the animal testing methods for dermal and ocular toxicity:

1. Incubation times *in vitro* are comparable to those used *in vivo*.
2. The test substance comes in direct contact with the eye epithelium.
3. With the exception of the uveal layer of the cornea, the lack of blood vessels allows for attainment of steady-state chemical concentrations in contact with tissue; thus results obtained from test protocols are not skewed by interruptions in absorption or local pharmacokinetic phenomena.

Alternatively, caution must be exercised with certain aspects of cell culture methodology when screening for ocular irritancy. For example, false positive and false negative results are usually errors of estimation and are closely related to the physicochemical properties of a test substance. Severely irritating properties of a substance *in vivo,* for instance, may be accounted for by extreme pH, which may not be detected *in vitro*. In addition, difficulty with solubility of a test substance in a solvent may underestimate true toxic concentrations.

While no single *in vitro* method accounts for all of these parameters, the combination of individual tests addresses these endpoints. The anticipation for success relies on diligent collective cooperation among the scientific, regulatory, industrial, commercial, and public representatives, using the concept of a battery of validated assays to screen chemicals with potential for human toxicity.

SUGGESTED READINGS

Barile, F.A., Continuous cell lines as a model for drug toxicity assessment, in *In Vitro Methods in Pharmaceutical Research*, Castell, J.V. and Gómez-Lechón, M.J., Eds., Academic Press, London, 1997.

Ekwall, B., Basal cytotoxicity data (BC data) in human risk assessment, in *Proceedings of Workshop on Risk Assessment and Risk Management of Toxic Chemicals*, National Institute for Environmental Studies, Ibaraki, Japan, 1992, p. 137.

Ekwall, B., The basal cytotoxicity concept, in *Alternative Methods in Toxicology and the Life Sciences*, Goldberg, A. and van Zupten, L.F.M., Eds., Mary Ann Liebert, New York, 1995, pp. 11, 721.

Ekwall, B. and Barile, F.A., Standardization and validation, in *Introduction to In Vitro Cytotoxicology: Mechanisms and Methods*, Barile, F.A., Ed., CRC Press, Boca Raton, FL, 1994, chap. 11.

Gautheron, P., Dukic, M., Alix, D., and Sina, J.F., Bovine corneal opacity and permeability test: an *in vitro* assay of ocular irritancy, *Fundam. Appl. Toxicol.*, 18, 442, 1992.

REVIEW ARTICLES

Atkinson, K.A., Fentem, J.H., Clothier, R.H., and Balls, M., Alternatives to ocular irritation testing in animals, *Lens Eye Toxic. Res.*, 9, 247, 1992.

Aufderheide, M., Knebel, J.W., and Ritter D., Novel approaches for studying pulmonary toxicity *in vitro*, *Toxicol. Lett.*, 140, 205, 2003.

Barile, F.A. and Cardona, M., Acute cytotoxicity testing with cultured human lung and dermal cells, *In Vitro Cell Develop. Biol.* 34, 631, 1998.

Barile, F.A., Arjun, S., and Hopkinson, D., *In vitro* cytotoxicity testing: biological and statistical significance, *Toxicol. in Vitro*, 7, 111, 1993.

Basketter, D.A. et al., The classification of skin irritants by human patch test, *Food Chem. Toxicol.*, 35, 845, 1997.

Bishop, A.E. and Rippon, H.J., Stem cells: potential for repairing damaged lungs and growing human lungs for transplant, *Expert Opin. Biol. Ther.*, 6, 751, 2006.

Borenfreund, E. and Puerner, J.A., Cytotoxicity of metals, metal–metal and metal–chelator combinations assayed *in vitro*, *Toxicology*, 39, 121, 1986.

Braun, A. et al., Cell cultures as tools in biopharmacy, *Eur. J. Pharm. Sci.*, 11, S51, 2000.

Bruner, L.H., Carr, G.J., Curren, R.D., and Chamberlain, M., Validation of alternative methods for toxicity testing, *Environ. Health Perspect.*, 106, 477, 1998.

Burton, A.B., York, M., and Lawrence, R.S., The *in vitro* assessment of severe eye irritants, *Food Cosmet. Toxicol.*, 19, 471, 1981.

Cockshott, A. et al., The local lymph node assay in practice: a current regulatory perspective, *Hum. Exp. Toxicol.*, 25, 387, 2006.

Dearman, R.J., Basketter, D.A., and Kimber, I., Local lymph node assay: use in hazard and risk assessment, *J. Appl. Toxicol.*, 19, 299, 1999.

Dolbeare, F. and Vanderlaan, M., Techniques for measuring cell proliferation, in *In Vitro Toxicity Indicators, Methods in Enzymology,* Tyson, C.A. and Frazier, J.M., Eds., Academic Press, San Diego, 1994, pp. 1B, 178.

Elaut, G. et al., Molecular mechanisms underlying the dedifferentiation process of isolated hepatocytes and their cultures, *Curr. Drug Metab.*, 7, 629, 2006.

Fentem, J.H. and Botham, P.A., ECVAM's activities in validating alternative tests for skin corrosion and irritation, *Altern. Lab. Anim.*, 30, 61, 2002.

Forbes, B. and Ehrhardt, C., Human respiratory epithelial cell culture for drug delivery applications, *Eur. J. Pharm. Biopharm.*, 60, 193, 2005.

Freed, L.E. and Vunjak-Novakovic, G., Space flight bioreactor studies of cells and tissues, *Adv. Space Biol. Med.*, 8, 177, 2002.

Ginis, I. et al., Differences between human and mouse embryonic stem cells, *Develop. Biol.*, 269, 360, 2004.

Gomez-Lechon, M.J., Donato, M.T., Castell, J.V., and Jover R., Human hepatocytes in primary culture: the choice to investigate drug metabolism in man, *Curr. Drug Metab.*, 5, 443, 2004.

Guha, C., Lee, S.W., Chowdhury, N.R., and Chowdhury, J.R., Cell culture models and animal models of viral hepatitis, Part II: hepatitis C, *Lab. Anim.*, 34, 39, 2005.

Hoffman, J. et al., Epidermal Skin Test 1000 (EST-1000): a new reconstructed epidermis for *in vitro* skin corrosivity testing, *Toxicol. in Vitro*, 19, 925, 2005.

Huggins, J., Alternatives to animal testing: research, trends, validation, regulatory acceptance, *ALTEX*, 20, 3, 2003.

Interagency Coordinating Committee on the Validation of Alternative Methods (ICCVAM), *Report of International Workshop on In Vitro Methods for Assessing Acute Systemic Toxicity*, NIH Publication 01-4499. National Institute of Environmental Health Sciences, Research Triangle Park, NC, 2001. http://iccvam.niehs.nih.gov/docs/guidelines/subguide.htm.

Interagency Coordinating Committee on the Validation of Alternative Methods (ICCVAM), *Guidance Document on Using in Vitro Data to Estimate in Vivo Starting Doses for Acute Toxicity*, NIH Publication 01-4500. National Institute of Environmental Health Sciences, Research Triangle Park, NC, 2001. http://iccvam.niehs.nih.gov/docs/guide lines/subguide.htm.

Interagency Coordinating Committee on the Validation of Alternative Methods (ICCVAM), *Biennial Progress Report of National Toxicology Program*, NIH Publication 04-4509, National Institute of Environmental Health Sciences, Research Triangle Park, NC, 2003. http://iccvam.niehs.nih.gov/about/annrpt/bienrpt044509.pdf.

Jones, B. and Stacey, G., Safety considerations for *in vitro* toxicology testing, *Cell Biol. Toxicol.*, 17, 247, 2001.

Jones, P.A. et al., Comparative evaluation of five *in vitro* tests for assessing the eye irritation potential of hair care products, *Altern. Lab. Anim.*, 29, 669, 2001.

Kimber, I. et al., The local lymph node assay and skin sensitization: a cut-down screen to reduce animal requirements? *Contact Derm.*, 54, 181, 2006.

Louekari, K., Sihvonen, K., Kuittinen, M., and Somnes, V., *In vitro* tests within the REACH information strategies. *Altern. Lab. Anim.*, 34, 377, 2006.

Mosmann, T., Rapid colorimetric assay for cellular growth and survival: application to proliferation and cytotoxicity assays, *J. Immunol. Meth.*, 65, 55, 1983.

Robinson, M.K. and Perkins, M.A., A strategy for skin irritation testing, *Am. J. Contact Derm.*, 13, 21, 2002.

Robinson, M.K. et al., Non-animal testing strategies for assessment of the skin corrosion and skin irritation potential of ingredients and finished products, *Food Chem. Toxicol.*, 40, 573, 2002.

Stobbe, J.L., Drake, K.D., and Maier, K.J., Comparison of *in vivo* (Draize method) and *in vitro* (Corrositex assay) dermal corrosion values for selected industrial chemicals, *Int. J. Toxicol.*, 22, 99, 2003.

Ubels, J.L., Ditlev, J.A., Clousing, D.P., and Casterton, P.L., Corneal permeability in a redesigned corneal holder for the bovine cornea opacity and permeability assay, *Toxicol. in Vitro*, 18, 853, 2004.

Ubels, J.L., Pruis, R.M., Sybesma, J.T., and Casterton, P.L., Corneal opacity, hydration and endothelial morphology in the bovine cornea opacity and permeability assay using reduced treatment times, *Toxicol. in Vitro*, 14, 379, 2000.

Vermeir, M. et al., Cell-based models to study hepatic drug metabolism and enzyme induction in humans, *Expert Opin. Drug Metab. Toxicol.*, 1, 75, 2005.

14 Toxicokinetic Studies *In Vitro*

14.1 INTRODUCTION

Toxicokinetics refers to the events surrounding the metabolic fate of a compound administered *in vivo* or *in vitro* and how this metabolism affects the concentration of the compound in the target organ. Organs, tissues, and cells are the target organs to which the chemical distributes. Toxicokinetics therefore refers to numerous events that affect the absorption of a toxic dose of a chemical in the gastrointestinal tract, through the skin, via the lungs, or across the mucous membranes. To whatever extent absorption occurs, the chemical is modified to a related compound, albeit more or less toxic. This metabolism occurs primarily in the liver, but also takes place in other metabolically active organs such as the lungs or kidneys.

Metabolism notwithstanding, and depending on its physicochemical properties, the chemical either circulates or is then distributed to and sequestered in various organs. Finally, the substance is eliminated either actively or passively through the extensive filtration processes of the kidney. Alternatively, the chemical is selectively eliminated through the lungs, skin, or intestinal tract.

Toxicokinetics contributes to the risk assessment estimation for the chemical based on calculated, potential exposure to humans or animals. *In vitro* methods participate in toxicokinetic modeling and risk assessment by supplementing data derived from *in vivo* studies. Not surprisingly, most toxicokinetic effects are measured in cell culture systems as relatively simple interactions between chemical substances and cell functions.

The following discussion introduces the concepts of kinetic distribution and metabolism of test compounds in *in vitro* systems and coordinates the information currently available with future studies. Chapter 3, titled "Toxicokinetics," describes the general principles of absorption, distribution, metabolism, and elimination.

14.2 METABOLIC STUDIES IN CELL CULTURE

The liver plays a key role in the detoxification of toxic substances but it can also transform a chemical into a metabolite more toxic than the parent compound. It is possible for the toxic metabolite to induce damage to the liver and other target organs. The underlying mechanism for liver injury from many chemicals whose targets are not specific for that organ may be explained through a toxicokinetic distribution.

In primary cultures of hepatocytes, it is possible to study both metabolite-induced liver toxicity and the pharmacological aspect of liver metabolism: the detoxification or activation of chemicals. Cells from rodents and humans are used and maintained in cultures either as organ cultures, monolayers, or isolated cells in suspension. The cells from mice and rats are obtained from organ perfusion of the liver, using collagenase to separate the cells rapidly and gently. Cytochrome P450 enzymes are functional and are maintained in these systems from a few hours to several days. Cultures of human hepatocytes are also realized with co-culture of feeder layers of other continuous cell lines. Recently, functional human hepatocytes have been maintained in a viable state after freezing — very important because liver metabolism is species-specific. Another development is the use of cell lines with preserved P450 activity instead of primary cultures of liver cells. Finally, several systems co-culturing hepatocytes and other target cells are used for toxico-kinetic modeling studies.

Among the different cell types of cell cultures described above, metabolic and hepatocellular transporting properties are studied with a large number of substances. Examples of specific research areas include (1) comparison of metabolizing capacities of liver cells and S9 mixtures, (2) cytochrome P450 activity preservation in hepatocytes, (3) species-specific metabolism, (4) the protective effects of reduced glutathione (GSH) in the presence of liver damage, and (5) the comparison of the metabolizing capacity of the liver and those of other organs such as kidneys, lungs, and intestine.

14.2.1 BIOTRANSFORMATION IN CELL CULTURE

14.2.1.1 Enzymatic Metabolism

Most cells *in vivo* have the capacity to enzymatically metabolize chemicals to less toxic, more water-soluble compounds, that are subsequently excreted through the urine. This capacity is clearly most pronounced in liver cells, but kidney, lung, intestine, and skin have extensive capacities for xenobiotic metabolism. Most other organs, such as blood cells, endothelial cells, muscles, and connective tissues have relatively low capacities for biotransformation. The main aspect of the biotransformation of chemicals by the most active cells is the ability to add functional groups (phase I reactions) that later are bound to endogenous substances such as glucuronic acid or GSH, resulting in increased water solubility of the metabolite (phase II biotransformation).

The most important phase I enzyme is cytochrome P450 oxygenase, inducible by such chemicals as phenobarbital. Examples of other more commonly found enzymes are epoxide hydrolase (phase I) and the transferases (phase II). It is important to note that the described biotransformation may result in more reactive (toxic) metabolites. Details of these reactions are discussed in Chapter 3.

Maintenance cultures of specific cells with high metabolic activity display biotransformation capacity, including significant P450 activity for several hours or days. Whether this short period is sufficient to mimic clinically relevant metabolism of all types of chemicals is questionable. Early passages of finite lines, especially of fetal cells, may display certain metabolic activities. Most other cultures of both finite and continuous cell lines, however, demonstrate low or immeasurable P450 levels that are sometimes coupled to hydrolase or glucuronyl transferase activity. The P450 activity found in some cell lines is increased by induction *in vitro*. Thus, few cell lines have some biotransformation capacity compared to the least competent cell types *in vivo*.

Primary cultures of rat hepatocytes are used as metabolic activation systems in a variety of toxicity assays. As mentioned above, freshly isolated cells are used to ensure metabolic integrity since continuous cultures lose metabolic capacity with time. Furthermore, some differences exist in the abilities of animal and human tissues to metabolize chemicals, although variations among individual donors may circumvent the intrinsic variations encountered between species. Consequently, primary cultures are suitable *in vitro* models for assessing biotransformation *in vitro*.

14.2.1.2 Use of S9 Mixture

Two methods are presently available to increase the capacity of biotransformation of cell cultures: (1) target cells are cultured on a feeder layer of more metabolically competent cells, for example, BHK21 cells irradiated to inhibit mitotic division, and (2) whole microsomes are added to the culture medium or preferably an ultracentrifuged fraction of homogenized liver from rats or mice pretreated with Arochlor 1254, thus containing P450 and other important microsomal enzymes. This fraction plus the NADPH co-factor is called the S9 mixture.

The first attempts to overcome the obstacle of limited metabolic capacity in short-term test systems were based on the observation that the metabolism required to activate many carcinogens to their active form was localized primarily with mixed function oxidases in the endoplasmic reticula of liver cells. The shortcomings of the Ames test conclusively demonstrated that the predictivity of the bacterial test for carcinogens substantially increased with the addition of a crude fraction of rat liver homogenate. These findings established the hepatic S9 fraction as the activation system of choice in most short-term testing protocols.

S9-reinforced cell line cultures offer moderate advantages over primary cultures of hepatocytes: (1) the system is flexible, easy to prepare and can be used with a variety of cell culture models, (2) drug-metabolizing enzymes are relatively stable in storage and are characterized biochemically prior to incorporation into the experiment, and (3) the system enhances phase I reactions such as the formation of reactive metabolites, but induces relative decreases of the biosynthetic phase II reactions since energy-rich phase II cofactors are not included in the mix.

Significant disadvantages often discourage the use of the mix: (1) S9-supplemented cultures require the addition of exogenous co-factors such as NADPH, (2) supplementation is cytotoxic to many systems and test results and enzyme activities per milligram of protein vary considerably, (3) the mixtures often exhibit technical variations in the enzymes, co-factors, exposure conditions, concentrations of the S9 in the mix, and enzyme inducers, and (4) the enzyme activity may not yield results that reflect the species, sex, and specificity of the organ of origin.

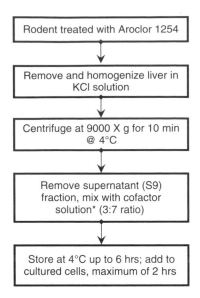

FIGURE 14.1 Protocol for preparation of S9 mixture. Co-factor solution consists of MgCl$_2$, glucose-6-phosphate, NADPH in phosphate buffer, 0.2 mol/L, pH 7.4. (*Source:* Flint, O.P. and Orton, T.C., *Toxicol. Appl. Pharmacol.*, 76, 383, 1984. With permission.)

14.2.1.3 Preparation of S9 Fractions

The S9 activation system is derived from the supernatant of the $9000 \times g$ spin from Aroclor 1254-induced rat liver homogenates. Upon injection in rats, Aroclor 1254* induces a broad spectrum of drug metabolizing enzymes. The liver is then homogenized, and the microsomes containing the phase I mixed function oxidase enzymes are isolated from the continuous endoplasmic reticulum fragments using differential centrifugation.

The supernatant S9 fraction is used as the crude homogenate or is further enriched for microsomal activity by additional high speed centrifugation. This separates a highly enriched microsomal pellet from the supernatant cytsolic fraction. The former contains phase I biotransformation enzymes while the supernatant is rich in many of the enzymes of phase II (conjugative) metabolism. An S9 buffer mixture is then formulated with co-factors as illustrated in Figure 14.1.

14.2.2 *IN VITRO* STUDIES OF ABSORPTION, DISTRIBUTION, METABOLISM, AND ELIMINATION (ADME)

The absorption, distribution, and elimination of a chemical *in vivo* are influenced by the interaction of biotransformation processes with the substance. The factors that affect these processes include:

---

* A mixture of polychlorinated biphenyls or a combination of phenobarbital and β-naphthaflavone is also used.

1. The ability and time that a substance requires to cross cellular barriers such as the intestinal mucous membranes, blood-brain barrier, and skin
2. The ability of the substance to enter and bind to the target site
3. The extent of binding to blood proteins and accumulation and sequestration in lipid compartments
4. The passage through the capillary endothelium
5. The water/lipid solubility ratio of the substance and its distribution through the intestine, blood, extracellular and intracellular fluids

The capacity to predict the mechanisms of toxicity and biotransformation of a chemical from its structural formula is feasible from structure–activity relationship models. It is also possible to predict absorption, distribution, and elimination from the structure of a substance and its reactivity with specific cellular transport proteins.

A chemical's physicochemical properties such as water and lipid solubility, ionization constant in acid or basic pH, and protein binding capacity are also instrumental in predicting its toxicokinetic characteristics. For example, large and highly ionized molecules have difficulty entering or traversing through cell membranes and are therefore poorly absorbed. Alternatively, small and moderately acidic or basic molecules are absorbed and distributed to the extra- and intracellular spaces. Moreover, highly lipophilic agents are readily absorbed and rapidly distributed to all physiologic compartments. In addition, they are often bound to tissues in high concentrations, and this may also account for the high degree of intracellular protein binding.

Most mutagenic and carcinogenic substances affect cells at relatively low concentrations. This is accounted for by biotransformation of the substance, and therefore is the reason most mutagens are classified as indirect acting. Substances that have general toxic effects, however, especially after acute toxic doses, demonstrate toxicity at higher concentrations that saturate the biotransforming capacity of the liver. This concentration is estimated if the dose and physicochemical properties of the substance are known. In summary, *in vitro* tests to determine distribution coefficient, protein binding, dissociation, and cellular transport are readily performed, and the physicochemical data generated from these studies supplement other types of *in vivo* toxicity data. Thus *in vitro* models for ADME are receiving increasing attention because they provide information for screening large numbers of lead compounds, especially when performed with high-throughput automation and analysis. These models are discussed in the literature cited below.

SUGGESTED READINGS

During, A. and Harrison, E.H., Intestinal absorption and metabolism of carotenoids: insights from cell culture, *Arch. Biochem. Biophys.*, 430, 77, 2004.

Flint, O.P. and Orton, T.C., An *in vitro* assay for teratogens with cultures of rat embryo midbrain and limb bud cells, *Toxicol. Appl. Pharmacol.*, 76, 383, 1984.

U.S. Food and Drug Administration, *Guidance for Industry-Drug Metabolism/Drug Interaction Studies in the Drug Development Process: Studies In Vitro*, Rockville, MD, 1997.

REVIEW ARTICLES

Andersen, M.E., Toxicokinetic modeling and its applications in chemical risk assessment, *Toxicol. Lett.*, 138, 9, 2003.

Bakand, S., Winder, C., Khalil, C., and Hayes, A., Toxicity assessment of industrial chemicals and airborne contaminants: transition from *in vivo* to *in vitro* test methods: a review, *Inhal. Toxicol.*, 17, 775, 2005.

Blaauboer, B.J., Biokinetic and toxicodynamic modelling and its role in toxicological research and risk assessment, *Altern. Lab. Anim.*, 31, 277, 2003.

Ding, S., Yao, D., Burchell, B., Wolf, C.R., and Friedberg, T., High levels of recombinant CYP3A4 expression in Chinese hamster ovary cells are modulated by co-expression of human reductase and hemin supplementation, *Arch. Biochem. Biophys.*, 348, 403, 1997.

Ekins, S. et al., Progress in predicting human ADME parameters *in silico*, *J. Pharmacol. Toxicol. Meth.*, 44, 251, 2000.

Gonzales, F.J. and Korzekwa, K.R., Cytochrome P450 expression systems, *Ann. Rev. Pharmacol. Toxicol.*, 35, 369, 1995.

Groneberg, D.A., Grosse-Siestrup, C., and Fischer, A., *In vitro* models to study hepatotoxicity, *Toxicol. Pathol.*, 30, 394, 2002.

Guillouzo, A., Liver cells models in *in vitro* toxicology, *Environ. Health Perspect.*, 106, 511, 1998.

Hayashi, Y., Designing *in vitro* assay systems for hazard characterization: basic strategies and related technical issues, *Exp. Toxicol. Pathol.*, 57, 227, 2005.

Kedderis, G.L. and Lipscomb, J.C., Application of *in vitro* biotransformation data and pharmacokinetic modeling to risk assessment, *Toxicol. Ind. Health*, 17, 315, 2001.

Lin, J.H., Tissue distribution and pharmacodynamics: a complicated relationship, *Curr. Drug Metab.*, 7, 39, 2006.

McCarley, K.D. and Bunge, A.L., Pharmacokinetic models of dermal absorption, *J. Pharm. Sci.*, 90, 1699, 2001.

Zucco, F., De Angelis, I., Testai, E., and Stammati, A., Toxicology investigations with cell culture systems: 20 years after, *Toxicol. in Vitro*, 18, 153, 2004.

15 Mutagenicity Testing *In Vitro*

15.1 INTRODUCTION

The last several decades have seen significant progress in the development of short-term tests for the detection of chemical mutagens. This advancement relies on understanding the complex processes leading to the potential carcinogenic actions of these chemicals and how short-term tests screen for human hazards. The *in vitro* tests encompass a wide variety of methods that are quicker, more economical, and more convenient than whole animal tests for the detection of mutations, chromosome breakage, and other genetic effects.

Mutations are changes in deoxyribonucleic acid (DNA) and ribonucleic (RNA) acid resulting in major consequences on normal cell proliferation, reproduction, physiology, and cell biology. Mutations exhibit immediate or delayed consequences but may also represent stable inherited changes in gene sequences resulting in phenotypic alterations. The type and extent of consequences of mutations depend on the dose, frequency, duration of exposure, and subsequent secondary effects caused by the response of the organism to the initial alteration. Several terms that are used to define the effect of chemicals on the chromosome apparatus include:

Mitogenesis — An immediate or delayed cellular response resulting in induction of increased cell proliferation

Genotoxicity — The effect of a chemical or its metabolites that induces changes to the genome and influences the production of malignant transformation

Chemical carcinogenicity — The effect of a chemical or its metabolites capable of inducing irreversible changes to normal cellular biology resulting in reversible or irreversible cellular transformation and uncontrolled cell proliferation

The changes result from genetic mutations, interfere with transfers of genetic information, and usually are manifested as lethal effects.

Considerable effort has been devoted to the development of short-term tests as methods for detecting, screening, and understanding the bases of chemical carcinogens. Short-term testing methods, therefore, are based on the accepted knowledge that mutations are the results in part of exposures to mutagenic and carcinogenic chemicals. Therefore, *in vitro* test systems use a variety of indicators that reflect mutational events in the presence of carcinogens and mutagens including microorganisms, plants, insects, and cultured mammalian cells to screen for potentially mutagenic, teratogenic, and carcinogenic agents. The advantages of *in vitro* tests depend on the rapid assessment of large numbers of chemicals, the suggestion of a mechanism for carcinogenicity or mutagenicity, the reduction, replacement, and refinement of animal testing, and a contribution to human and animal risk assessment that correlates as well as the predictive ability of animal toxicology testing.

15.2 BACTERIAL CELL SYSTEMS

Short-term tests use both prokaryotic and eukaryotic cells to detect the vast majority of potential carcinogens and mutagens. The prokaryotes include bacteria and blue-green algae (cyanobacteria). The bacterial test systems use *auxotrophic* organisms whose growth and proliferation depend on the presence of an essential rate-limiting nutrient component in the growth medium, usually an amino acid. The auxotrophic organism thus represents a mutant bacterium with a highly specific defect in a gene locus. In contrast, the normal or wild-type prototrophic organism lacks the mutation; its growth is not dependent on the addition of the essential rate-limiting nutrient. Prototrophic organisms are capable of maintaining normal cell proliferation from the minimal essential growth medium components. Figure 15.1 outlines typical growth properties of auxotrophic and prototrophic bacteria as preliminary experiments in preparation for the Ames test.

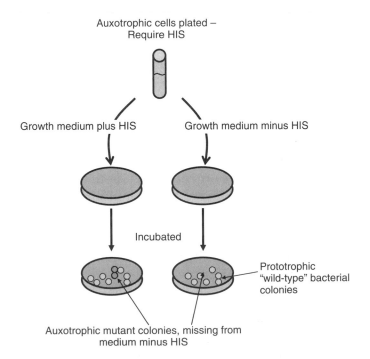

FIGURE 15.1 Growth properties of auxotrophic and prototrophic bacteria.

The foundation of the bacterial tests is based on (1) induction of a reverse mutation from the component-dependent strain to a wild-type capable of sustaining growth on minimal essential medium, and (2) induction of a forward mutation in which additional agent-induced phenotypic genetic changes are identified.

Detection of forward mutations results in larger genetic targets for a chemical, presenting several loci within specific genomes for identification of phenotypic changes. Reverse mutation assays provide specific and selective target sites for chemical action, relying on a second agent-induced mutation to modify the effect of a preexisting mutation.

15.2.1 Ames Test

The bacterial mutagenicity assay for a direct-acting mutagen the *reverse-mutation plate test*, *plate-incorporation test*, or *agar overlay test* uses bacterial strains mixed with known concentrations of a test agent. The suspension is incubated with agar solution containing the essential component for growth of the auxotrophic organism. This permits the bacterium to multiply through several generations in the presence of the test chemical. The mixture is then overlaid onto the surface of an agar plate lacking the minimum essential medium component (minimal agar).

On the agar plate, all auxotrophic organisms stop growing except those affected by the test agent that revert (back-mutate) to prototrophic growth patterns. Thus, mutated organisms, initially unable to grow in the absence of the rate-limiting essential component, consequently form progenies of a single revertant, "wild-type" bacterium upon depletion of the essential nutrient. Unless evidence of extensive genetic changes or lethal damage to the auxotrophic bacterial genome appears, the number of revertants is proportional to the concentration of test agent in the incubation medium. Figure 15.2 is a diagram of the Ames test and the growth properties of auxotrophic bacteria in the absence or presence of a mutagen.

Bacterial mutagenesis assays are the most widely used short-term tests for screening of potential mutagens and carcinogens. They are highly sensitive for genotoxic agents, are technically easy, fast, and economical. The most extensively used bacterial assay system uses mutant tester strains of *Salmonella typhimurium* developed by Ames (1972). The original method used strains that were genetically defective for their ability to synthesize histidine (HIS). Assay sensitivity was later improved by incorporating additional mutations in tester strains (Table 15.1) specifically designed to detect frame shift or base pair substitution mutations induced by different classes of mutagens.

All strains contain some type of mutation in the histidine operon, while additional mutations increase their sensitivity to mutagens. For instance, the *rfa* mutation causes partial loss of the lipopolysaccharide surface coatings of bacteria and thus increases permeability to large molecules that do not normally penetrate cell walls. Another mutation (*uvrB*) greatly increases the sensitivity of the bacteria to mutagens by deleting the gene coding for DNA excision repair.

Many of the standard tester strains also contain the R-factor plasmid, pKM101. These strains are sensitive to a number of mutagens that are inactive in non-R-factor parent strains. Thus, sensitivity for identifying weak mutagenic agents as well as

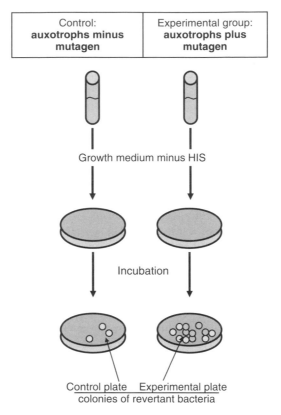

FIGURE 15.2 Ames test and growth properties of auxotrophic bacteria in presence or absence of a mutagen. Two cultures of HIS-dependent Salmonella strain are prepared. Suspected mutagen is added to the experimental group only. All samples are plated onto their respective agar plates in medium lacking HIS. Plates are incubated for 24 to 48 hr at 37°C. Bacteria that mutate back (revertants) to HIS-independence will form colonies. Colonies are counted and compared to spontaneous HIS-independent control plates.

TABLE 15.1
Incorporation of Mutations of Original Salmonella Strains Incurring Increased Sensitivity of Ames Test to Mutagens

Mutation	Feature
uvrB	Eliminates DNA excision repair system
rfa	Interferes with synthesis of LPS components of cell walls resulting in increased permeability of bacterial cell walls to mutagens

Note: LPS = lipopolysaccharide.

TABLE 15.2
Tester Strains Incorporating Additional Mutations

Tester Strain	Mutation of HIS Gene	Additional Mutations	Features
TA1535	Base pair substitutions	rfa, uvrB	Increased sensitivity to chemicals that interfere with base pair substitutions
TA100	Base pair substitutions	rfa, uvrB, plasmid pKM101	Features of TA 1535 + transfer antibiotic resistance, R factor plasmid, to TA1535
TA1537	Frame shift mutations	rfa, uvrB, plasmid pKM101	Increased sensitivity to chemicals that interfere with frameshift mutations
TA 1538	Frame shift mutations	rfa, uvrB	Increased sensitivity to chemicals that interfere with frameshift mutations
TA 98	Frame shift mutations	rfa, uvrB, plasmid pKM101	Features of TA 1538 + transfer antibiotic resistance, R factor plasmid, to TA1538

specificity for detecting different classes of carcinogens was a significant advancement in the development of the test (Table 15.2). Today, the Ames test is the only short-term assay that is validated, popular, and sensitive for detecting carcinogens. Its flexibility allows its adaptation to testing large numbers of chemicals with varying characteristics. Nevertheless, awareness of its limitations is important in evaluation and interpretation of data. Probably the most important consideration is adherence to standardized protocols to limit inter- and intra-laboratory variability. Two variations of the Ames, the spot test and the standard text, are described below.

15.2.1.1   Spot Test

The spot test is the simplest technique for rapid screening of large numbers of chemicals. TA98, TA100, or TA1535 *Salmonella typhimurium* tester strains are incubated in minimal medium. The organism is applied with an agar overlay containing a rate-limiting quantity of histidine to initiate growth (plate incorporation test). A filter disk soaked with known concentrations of the test agent is placed and incubated on the agar. The toxicity of the diffusible solution containing the dissolved chemical is proportional to the diameter of the zone of growth inhibition versus the growth of revertant bacteria around the application site. Although water-soluble agents do not easily penetrate the agar and dilution of the chemical concentration increases with outward diffusion, the test is useful in detecting a dosage range of mutagens. The procedure is constructed to detect several sensitive and insensitive tester strains in each agar plate.

15.2.1.2   Standard Ames Test

As briefly noted above, the original test system used known concentrations of tester strains incubated with liquid agar, histidine, and increasing concentrations of the test chemical. The mixture was applied to a petri dish containing minimal medium agar (minus HIS), incubated for 48 to 72 hr, and the revertant colonies were counted against control groups (and spontaneous revertants). The principal limitations of the test were the expectation that the assay could screen for all classes of chemical carcinogenicity and that the bacterial system could mimic the metabolic capacities of mammalian organisms to activate pro-carcinogens chemicals that necessitate metabolic activation to the more active species.

Thus, some important classes of compounds such as environmental carcinogens, polycyclic aromatic hydrocarbons and nitrosamines exhibit minimal mutagenic activity in the absence of exogenous metabolic activation. The solution to the dilemma resided in the discovery that homogenates of mammalian liver provided the source of cytochrome P-450 mixed function oxidases (MFOs). These enzymes present in the endoplasmic reticula (ER) of hepatocytes are useful for converting pro-carcinogenic agents into active carcinogens. Thus, the predictivity of the bacterial test system for carcinogens was substantially improved with the addition of rat or human hepatic homogenates to the incubation mixture. In addition, it established the hepatic S9 fraction as the preferred activation system (see Chapter 14) and solidified the Salmonella–mammalian microsome test method as the choice for detecting short-term genotoxic agents.

In order to obviate the highly variable activity levels of hepatic enzymes present among specimens, young male rats are treated with Aroclor 1254 (a mixture of polychlorinated biphenyls, PCBs). After 5 days, the liver homogenates are processed by differential centrifugation to exclude nuclei, most organelles, and membrane fragments. The S9 supernatant fraction is isolated from continuous ER fragments that form microsomes containing high concentrations of phase I MFOs. The fraction is used as a crude homogenate, is further enriched by additional high speed centrifugation for phase II enzymes, or is frozen.

To complete the short-term test, an aliquot of the S9 fraction is added to the Salmonella–test chemical–agar buffered suspension, overlayed onto a minimal medium agar plate, and incubated for 48 to 72 hr at 37°C. The revertant colonies are counted against controls (Figure 15.2).

15.3 MAMMALIAN CELL SYSTEMS

15.3.1 Basis of Mutagenicity Testing with Mammalian Systems

Determining mutagenic profiles of chemicals such as pesticides and herbicides requires tests of high sensitivity and specificity. An effective strategy uses tests that produce reproducible and biologically relevant data based upon three stages. Stage 1 uses bacterial gene mutation assays as described above; stage 2 assays are cytogenetic tests that monitor clastogenicity and aneugenicity (described later in this chapter); and, stage 3 assays record the induction of gene mutations in cultured mammalian cells.

Mutagenesis testing in mammalian cells is conducted as part of stage 3 to confirm that a chemical is presumed mutagenic for higher animals. Testing in mammals supposes that higher degrees of DNA repair and xenobiotic metabolism are markedly different from those of bacterial systems. Consequently, several assays incorporate various continuous cell lines and primary cultures of animal and human origin. While primary cultures maintain some biotransformation capability, continuous cell lines have limited potential. Cell culture systems thus permit a closer examination of the mechanisms of action responsible for gene mutation and chromosome aberrations.

Short-term mammalian tests detect mutations at the hypoxanthine–guanine phosphoribosyl transferase (HGPRT) locus using Chinese hamster lung (V79) or ovary (CHO) cells; mutations at the thymidine kinase (TK) locus, in the mouse lymphoma L5178Y cell line; or the TK6 human lymphoblastoid line. Chemical treatment of the cells induces a forward mutation at a specific locus, after which the cultures are incubated to allow phenotypic expression of the induced mutants. The mutants form colonies in medium containing a selective agent that allows only mutants to proliferate while inhibiting wild-type cells. Mutagenic potential is proportional to mutant cell colony formation.

V79, CHO, L5178Y, and TK6 cells lack the metabolic capacity to activate promutagens and are therefore supplemented with an exogenous activation system including hepatic S9 fractions or intact cellular activation systems. The addition of exogenous factors and systems, however, introduces a major source of variability within the testing system.

15.3.2 Cell Lines

15.3.2.1 CHO and V79 Cells

Chinese hamster ovary (CHO) and pulmonary (V79) cells have stable karyotypes (2N = 20, 21, respectively), with a doubling time of 12 to 16 hr. The continuous cell lines are used with a variety of genetic markers including resistance to 8-azaadenine, cycloheximide, and methotrexate. The mutagenesis system involves a mutation at the HGPRT locus, the gene of which is present on the X chromosome. The locus encodes for the HGPRT enzyme responsible for the conversion of the purines, hypoxanthine and guanine, into nucleotides. Chemical agents such as 6-thioguanine (6-TG) and 8-azaguanine are cytotoxic to cells possessing HGPRT (HGPRT+), whereas a mutation at the HGPRT locus (HGPRT−) induces resistance.

The assay relies on the metabolism of 6-TG and 8-azaguanine by HGPRT to toxic ribophosphorylated derivatives that cause cell death when incorporated into DNA. CHO or V79 cells (HGPRT+) are incubated with the test agent in the presence or absence of S9 supplementation. The cultures are incubated and replated in medium containing 6-TG. Mutations at the HGPRT locus (HGPRT−) inactivate the enzyme and inhibit the catalytic formation of toxic metabolites from 6-TG. The mutant cells are able to survive using *de novo* purine biosynthesis and form colonies at concentrations of 6-TG normally cytotoxic to wild-type cells.*

---

* The CHO systems are also used extensively to examine chromosomal aberrations and sister chromatid exchanges, thus allowing comparison of data from tests having different endpoints.

15.3.2.2 Mouse Lymphoma L5178Y Cells

The mouse lymphoma L5178Y mutagenesis system is based on quantifying forward mutations occurring at the heterozygous TK locus (L5178Y/TK±). The gene expresses the TK enzyme responsible for the incorporation of exogenous thymidine monophosphate. When trifluorothymidine (TFT) is substituted for thymidine as the substrate, the resultant phosphorylated TFT irreversibly inhibits thymidylate synthetase, resulting in cell death. Mutations at the locus result in the loss of thymidine kinase activity and the formation of a homozygous strain (TK–/–). Incorporation of 5-bromodeoxyuridine (BrdUrd) results in cytotoxicity to the TK± cells, while the enzyme deficient cells (TK–/–) continue proliferation as they are unable to convert BrdUrd to toxic metabolites.

When lymphoma cells are exposed to a putative mutagen in the L5178Y mutagenesis system, forward mutations occur at the TK locus and inactivate the TK enzyme. As with the CHO and V79 cells, the inclusion of TFT in the culture medium selects for TK mutants that proliferate by synthesizing purines *de novo*, whereas wild-type cells do not survive.*

15.3.3 CYTOGENETIC TESTING

The second stage of tier testing for the detection of mutagenicity and carcinogenicity includes the cytogenetic tests. Three different types of *in vivo* and *in vitro* cytogenetic tests monitor clastogenicity and aneugenicity. These light microscopic tests are used to determine chemically induced changes in a structure or number of chromosomes and include tests to determine (1) chromosome aberrations, (2) exchanges of chromosome materials among sister chromatids, and (3) chromosome fragments or micronuclei arising from chromosome damage.

Originally developed for *in vivo* testing, advances in cell and tissue culture technology have created advantages for *in vitro* protocols. In particular, the cell culture adaptations are more convenient, economical, and are capable of screening many compounds. In fact, some procedures incorporate sequential use of both *in vivo* and *in vitro* methods, thus allowing more applicability for human exposure to genotoxic agents.

15.3.3.1 Chromosome Aberrations

Congenital abnormalities and neoplastic growth have long been considered to result in part from chromosome aberrations. Most known carcinogens are also identified as *clastogens* — agents that induce chromosome breakage. As described in Chapter 6 "Acute Toxicology Testing," *in vivo* methods involve microscopic examinations of different cell types from animals previously treated with a test agent. In contrast, *in vitro* methods are more sensitive but, as with bacterial mutagenic systems, usually require exogenous supplementation for enzymatic activation. Nevertheless, CHO

---

* Interestingly, the TK mutagenesis assay is also useful for detecting the clastogenic potential of a test compound, that is, the induction of chromosome breakage.

and V79 cell lines and human lymphocytes are commonly used for short-term cytogenetic studies.

Protocols are similar for most cell types and are similar to the protocol described for karyotypic analysis (Chapter 12, Cell Culture Methodology). The procedure commences with the induction of proliferation of cells, when needed, to stimulate entry into the cell cycle. This is accomplished by the addition of a mitogen such as phytohemagglutinin (PHA). Initial dosage range determination of the test chemical includes measurement of cell viability or mitotic index as a baseline control parameter for the cell line. Sub-confluent cultures in logarithmic phase of growth are prepared in LabTek™ chambers and treated for an empirically determined period to encompass all stages of the cell cycle.* The cultures are washed; colchicine is added to arrest the cells in metaphase and the cultures are processed 2 to 3 hr later for analysis of chromosome aberrations. Figure 15.3 illustrates the principles of typical models of chromosome aberration, exchange, and reunion.

A metaphase spread relies on proper swelling of the cells by the addition of hypotonic potassium chloride solution. Cells are then fixed, the chamber cover is removed to expose the fixed cells on the plastic or glass slide, and the slide is stained with methylene blue or Giemsa counterstain. Chromosomes are examined microscopically (1000×, oil immersion), and numbers and types of aberrations are scored.

15.3.3.2 Sister Chromatid Exchange

Sister chromatid exchange (SCE) involves the cross-over of replicating DNA between chromatids at homologous loci. Unlike chromosome aberration, the interchange does not involve morphological alteration and is detected only by differential labeling of the sister chromatids (Figure 15.3B). Traditional methods using [³H]-thymidine ([³H]-TdR) incorporation into newly transcribed DNA have been replaced by methods using BrdUrd labels combined with different stains.

The incorporation of BrdUrd into cells offers a more accurate estimation of the fractions of cells in S phase than does DNA measurement alone using [³H]-TdR uptake. The optimal method is the simultaneous measurement of both DNA content and incorporated BrdUrd into cells. Continuous exposure to BrdUrd label in a method analogous to those developed for [³H]-TdR provides an estimate of the fraction of non-cycling cells, and pulse-chase experiments provide quantitative measures of cell cycle progression. For instance, pulse-chase exposure to BrdUrd (0.5 to 2 hr) labels that fraction of cells in S phase at the time of the pulse, whereas continuous exposure (more than 2 hr) tags all cells that pass through S phase during the exposure period.

The protocol is similar to karyotypic analysis and detection of chromosome aberrations, with the exception that the BrdUrd label is added to the culture medium for at least two population doubling levels (PDLs). Cells are grown in LabTek chambers; pulse-chase and continuous exposures to the halopyrimidine are used after incubation with the test chemicals. After appropriate wash steps, sister chro-

---

* In the presence of S9 supplementation, chemical exposures are reduced to 2 to 3 hr due to cytotoxicity of the additive.

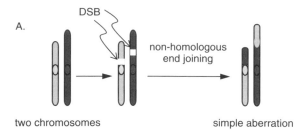

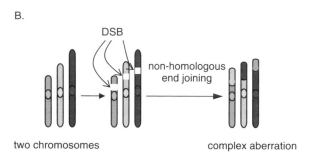

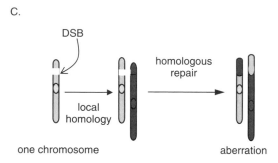

FIGURE 15.3 Standard models of chromosome aberrations. A = simple aberration, two double-strand break (DSB) misjoin. B = rejoining of more than two DSBs to form complex aberration. C = recombinational misrepair where a single-strand break leads to enzymatically mediated homologous misrepair.

matids are distinguished with a fluorescent label (anti-BrdUrd IgG-FITC conjugated antibody) plus Giemsa stain (FPG method) that produces permanently and differentially stained chromosomes.* Blocking controls and biological controls are performed in parallel sets to account for background non-specific fluorescence and autofluorescence, especially that observed with G_1 cells that tend to glow with minimal green fluorescence from antibodies to BrdUrd.

---

* Fluorescence has the advantage of determining labeling index defined as BrdUrd labeled cells per total cells observed in the field.

It is commonly accepted that a strong correlation exists between the induction of SCE and carcinogenicity. In addition, SCE testing is considerably less troublesome than tests for chromosome aberrations, thus validating SCE detection as a dominant and useful short-term test for carcinogens and genotoxic agents.

15.3.3.3 Detection of Micronuclei

Tests designed for the measurement of micronuclei are comparable in sensitivity and speed to detection of chromosome aberrations, although the former tests detect only chromosome breakage. Micronuclei arise from chromosome fragments that are not successfully passed to daughter nuclei during mitotic cell division, particularly due to the lack of a centromere.

The assays are also used with a variety of plant and mammalian cells in culture; proliferating cells are particularly useful. Slight variations in procedure from those used for chromosome aberrations are noted in the protocols for the detection of micronuclei. Briefly, cultured cells are treated with a test agent for a selected period, then incubated in the presence of cytochalasin B to prevent cytokinesis*; nuclear division is not blocked (colchicine is omitted from the procedure). The rest of the protocol proceeds as for chromosome aberrations. Cells that contain two nuclei are scored for the presence of micronuclei that are not incorporated into daughter nuclei.

15.3.3.4 Limitations of Short-Term Cytogenetic Testing

As mentioned above, cultured cells lack the metabolic potential of biotransformation, thus limiting measurement of genotoxicity to non-metabolized parent chemicals. Microsomal enzyme preparations, feeder layers, and co-cultures, however, may supplement for the lack of metabolizing enzymes. While the test for chromosome aberrations is more sensitive and dependable, it is more laborious and more expensive than the micronucleus test or the SCE test.

15.3.4 Unscheduled DNA Synthesis (UDS)

15.3.4.1 Principles of Testing

As mammalian cells both *in vivo* and *in vitro* are continuously insulted with natural and experimental exposures to chemicals, they have by evolutionary selection developed defense systems that facilitate repair of the inflicted damage. The repair mechanism of the process of unscheduled DNA synthesis (UDS) is well characterized. The principles of this toxicology test rely on the incorporation of [^{3}H]-TdR into cultured cells during or after treatment with the test chemical. The cell preparations are fixed and dried on microscope slides and exposed to a phosphorescent digital imaging system.† Other methods involve quantitative measurement of uptake of [3H]-TdR into newly synthesized or repaired DNA from control and chemically treated groups.

---

* Cytoplasmic division.
† Traditional autoradiographic methods involved exposure of slides to a photographic emulsion that usually required several days to visualize opaque radioactive spots in nuclei.

UDS estimation is an indirect measurement of DNA damage, since the sequential enzymatic repair process is initiated by a cell's recognition of the presence of DNA adducts formed as a result of chemical insult. The primary DNA structure is restored through excision of the adducts, DNA strand polymerization, and subsequent ligation. The intranuclear incorporation of the radioactive nucleotide base during the repair process acts as a quantitative tracer for the repaired lesion during any period of anticipated or unanticipated (unscheduled) DNA synthesis. The assay generates semi-quantitative or quantitative data by autoradiography or liquid scintillation counting, respectively.

15.3.5 Cell Transformation

15.3.5.1 Principles of Testing

Short-term tests for cell transformation represent sensitive and mechanistically valid *in vitro* models for chemically induced cancers *in vivo*. Malignant transformation is accompanied by phenotypic alterations that are retrospectively related to carcinogenic induction. The tests use a variety of endpoints that rely on the apparent alterations to detect the transforming abilities and carcinogenic potentials of test compounds, including (1) loss of anchorage dependence, and (2) alterations cellular morphology or growth characteristics.

The latter criterion is measurable using typical cell culture parameters such as loss of contact inhibition, altered growth in agar, appearance of immortal characteristics, proliferation in multilayer cultures, and viral dependence. In addition, transformed cells acquire the ability to induce cancer when transplanted into syngeneic immunologically deficient animals. The tests however suffer from lack of reproducibility, tremendous variability, and subjective interpretation encountered with the scoring of some results.

15.3.5.2 Focus Transformation Assay

Immortal cell lines, C3H/1OT1/2 and BALB/c 3T3 fibroblasts, are used in focus transformation assays. The cells display characteristics of immortal mammalian cultures, particularly their continuous proliferation and tumorigenic propensity *in vivo*. Unlike other human or rodent immortal cells such as A549 cells derived from human lung carcinomas and Caco-2 cells from human colon carcinomas, C3H/1OT1/2 and BALB/C 3T3 organize into a spindle shape morphology and exhibit density-dependent and contact inhibition of growth. In addition, they form foci in monolayer cultures.

The assay consists of exposing cultures to test chemicals for 24 to 72 hr upon initial plating, after which an additional 4 to 6 wk of passage and exposure to chemicals expresses the transformed phenotype (Figure 15.4). Monolayers or multilayers are fixed and stained to determine the presence of foci of transformed cells. Variability of results arises from the subjective influence of scoring. Thus, attempts to standardize the assay are based on the recognition of foci according to the type observed:

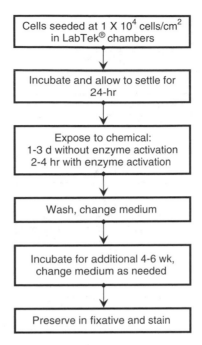

```
┌─────────────────────────────────┐
│ Cells seeded at 1 X 10⁴ cells/cm²│
│      in LabTek® chambers         │
└─────────────────────────────────┘
                 │
                 ▼
┌─────────────────────────────────┐
│ Incubate and allow to settle for │
│              24-hr               │
└─────────────────────────────────┘
                 │
                 ▼
┌─────────────────────────────────┐
│        Expose to chemical:       │
│   1-3 d without enzyme activation│
│    2-4 hr with enzyme activation │
└─────────────────────────────────┘
                 │
                 ▼
┌─────────────────────────────────┐
│        Wash, change medium       │
└─────────────────────────────────┘
                 │
                 ▼
┌─────────────────────────────────┐
│  Incubate for additional 4-6 wk, │
│     change medium as needed      │
└─────────────────────────────────┘
                 │
                 ▼
┌─────────────────────────────────┐
│    Preserve in fixative and stain│
└─────────────────────────────────┘
```

FIGURE 15.4 Outline of focus transformation assay using C3H/1OT1/2 and BALB/c 3T3 fibroblasts.

- Type I foci are not scored as malignantly transformed; they are tightly packed foci and monolayer cultures are not tumorigenic after inoculation into host animals.
- Type II foci exhibit dense, overlapping, multilayer networks of cells.
- Type III foci are densely stained overlapping groups of multilayers of cells; the edges of the growth patterns are irregular and are tumorigenic upon injection into host animals.

The assay has particular value for understanding and screening mechanistic initiation and promotion of multistage carcinogens that occur *in vivo*.

15.3.5.3 Syrian Hamster Embryo (SHE) Cell Transformation Assay

This assay incorporates the mortal cell line into both focus and clonal assays. The focus assay begins with target SHE cells seeded at 1×10^5 cells/cm² in the absence of feeder layers. This represents a higher seeding density since normal cell inoculation densities for most mammalian cells requires about 10-fold fewer cells. The cells are incubated for 3 days followed by a chemical exposure for 3 days, and transformed foci are evaluated after an additional 20 to 25 days.

The clonal assay features initial plating of mitotically inactivated feeder layers up to the confluent density and co-cultured with a few hundred target SHE cells.

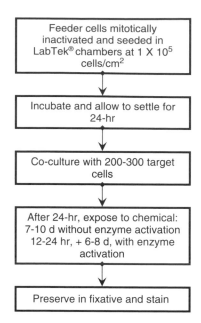

FIGURE 15.5 Outline of Syrian hamster embryo (SHE) cell transformation assay using SHE fibroblasts. Cells are mitotically inactivated using either ultraviolet light or mitomycin-C.

After 24 hr, the cultures are exposed to the test chemical for 7 to 10 days and processed for histological transformation. SHE cells transform into condensed, separate colonies with clearly demarcated boundaries in contrast to the appearance of traditional cell transformation assays. Figure 15.5 outlines the procedure.

15.3.5.4 Viral and Chemical Transformation Methods

Historically applied in mechanistic carcinogenic toxicology studies, viral and chemical transformation assays have also found some utility as short-term screening tests. Two well categorized transformation methods include SHE/SA7 cell lines and Fisher rat embryo (FRE) cells (FRE/RLV transformation assay). Both cell lines are more susceptible to chemical transformation since the cells were previously exposed to infection with simian adenovirus (SA7) and with Rauscher leukemia virus (RLV), respectively.

15.4 DEVELOPMENT OF SHORT-TERM TEST STRATEGIES

Important objectives in the development of *in vitro* short-term tests for the evaluation of mutagenic potential of chemicals include the identification of potential germ cell and somatic cell mutagens as potential carcinogens. With the current understanding and developmental status of *in vitro* methods, it is unreasonable to assume that even a battery of *in vitro* tests alone can meet these objectives, especially for human risk assessment. Consequently, evaluation of mutagenic potential of chemicals, especially chemicals with direct human exposure, necessarily should incorporate multiple assays, the objective of which is to identify genotoxic and cytogenetic carcinogens at an early stage. Thus, the ability of any testing scheme to detect potential carcinogens involves a variety of methods, including:

1. A preliminary screen using structure-activity relationship (SAR) studies of the chemical
2. *In vitro* tests genetic tests with bacterial cells and chromosome tests with mammalian cells
3. *In vivo* assays select carcinogenicity studies with animal models based on the preliminary screens

The performance of screening tests for the detection of mutagenic potential of chemicals is guided by the quantitative ability of the tests for sensitivity, specificity, and concordance. Sensitivity is the percentage of carcinogens that screen positive in a given test; specificity is the percentage of non-carcinogens that are negative; concordance refers to the percentage of results from both carcinogens and non-carcinogens from short-term tests that agree with *in vivo* data.

The European Scientific Committee on Cosmetics and Non-Food Products (SCCNFP) guidelines for testing of hair dyes for genotoxic, mutagenic, and carcinogenic potential include a battery of six *in vitro* tests. The guidelines suggest that potential genotoxic activity of chemicals used in hair dyes may effectively be determined by the application of a limited number of well validated test systems capable of detecting induced gene mutations and structural and numerical chromosome changes.

The recently updated guidelines for food additives by the European Community's Scientific Committee on Food recommends a battery of three *in vitro* tests for the induction of gene mutations in bacteria and in mammalian cells (mouse lymphoma TK assay and assay for chromosome aberrations). Also, additional data generated from physicochemical, structural, and metabolic properties may supplement the standard battery.

Several important international programs are directed at harmonization of genetic toxicology testing strategies and testing standards. The Organization for Economic Cooperation and Development (OECD) has a major role in developing recommendations for internationally harmonized testing protocols. The protocols developed by OECD are updated periodically (http://www.oecd.org/). Table 15.3 lists OECD guidelines on genetic toxicology testing and guidance on the selection and application of *in vitro* assays.

The EU also describes a number of tests similar to the OECD guidelines (http://europa.eu.int/) that include the morphological cell transformation assay. Among the national guidelines of the EU, those of the United Kingdom Department of Health are particularly noteworthy. The "Guidelines for the Testing of Chemicals for Mutagenicity" (http://www.doh.gov.uk/com.htm) are based on recommendations of the Committee on Mutagenicity of Chemicals in Food, Consumer Products

TABLE 15.3
OECD Guidelines on *In Vitro* Genetic Toxicology Testing and Guidance on the Selection and Application of Assays

Guideline No.	Title	Original Adoption Date	Most Recent Update
471	Bacterial reverse mutation test	1983	1997
473	In vitro mammalian chromosome aberration test	1983	1997
474	Mammalian erythrocyte micronucleus test	1983	1997
475	Mammalian bone marrow chromosome aberration test	1984	1997
476	In vitro mammalian cell gene mutation test	1984	1997
477	Genetic toxicology: sex-linked recessive lethal test in Drosophilia melanogaster	1984	None
478	Genetic Toxicology: Rodent Dominant Lethal Test	1984	None
479	Genetic toxicology: in vitro sister chromatid exchange assay in mammalian cells	1986	None
480	Genetic toxicology: gene mutation assay in Saccharomyces cerevisiae	1986	None
481	Genetic toxicology: mitotic recombination in Saccharomyces cerevisiae	1986	None
482	Genetic toxicology: DNA damage and repair, unscheduled DNA synthesis in mammalian cells	1986	None
483	Mammalian spermatagonial chromosome aberration test	1986	1997
484	Genetic toxicology: mouse spot test	1986	None
485	Genetic toxicology: mouse heritable translocation assay	1986	None
486	Unscheduled DNA synthesis test with mammalian liver cells	1997	None

Source: OECD Genetic Toxicology Testing Summaries: http://www.oecd.org

and the Environment (COM) and provide background information and justification for testing and assessment. The proposal extends to recommend the inclusion of an *in vitro* test for the potential to induce aneuploidy (micronucleus test or metaphase analysis).

The International Conference on Harmonization (ICH) of the Toxicological Requirements for Registration of Pharmaceuticals for Human Use recommends a core testing battery for evaluation of pharmaceuticals. The working group suggested that a battery include a test for gene mutation in bacteria for chromosome aberrations in mammalian cells *in vitro* or the L5178Y mouse lymphoma mammalian cell mutagenesis test and an *in vivo* test for chromosome damage in rodent hematopoietic cells. The ICH guidance documents address test procedures as well as strategy and test interpretation.

Reviews of genotoxicity testing programs and publications are documented as part of the GENETOX Program of the U.S. Environmental Protection Agency (EPA) (http://toxnet.nlm.nih.gov/cgi-bin/sis/htmlgen?GENETOX). The recently completed phases of its ongoing efforts to review the existing literature about chemically induced genetic toxicology were derived from peer-reviewed mutagenicity test data. The agency also gives direction to the diversity of the tests used to detect chemically induced genetic damage.

EPA's Environmental Carcinogenesis Division (ECD) is located within the National Health and Environmental Effects Research Laboratory (NHEERL). Its research focuses on improved understanding of environmentally induced mutagenesis and carcinogenesis for incorporation into human cancer risk assessment models. The program uses cellular, animal, and computer models for assessing responses to a broad range of environmental chemicals and mixtures. The objective of the research is to understand the chemical induction of somatic and germ cell mutations as a basis for improving risk assessments for cancer and heritable mutations, respectively. Goals and approaches of the research are directed at:

1. Identifying hazards
2. Assessing dose and tumor responses using applicable biomarkers
3. Developing dose response curves for tumor outcomes in humans using rodent tumor data and cellular indicators of tumorigenicity
4. Defining mechanisms of metabolic activation and detoxification for mutagenic agents
5. Enabling comparison of human and laboratory animal and cellular test system results
6. Improving the performance of biologically based dose response models used in risk assessment
7. Developing and validating molecular techniques, short-term bioassays, and whole animal bioassays for evaluating carcinogenic potential

Thus, ECD has the capability to influence the development of the cancer risk assessment process both nationally and globally.

15.5 SUMMARY

A variety of factors contribute to the heterogeneous results obtained from *in vitro* mutagenicity tests, the most important of which include:

1. Absence of cellular biotransformation systems in prokaryotic cells
2. Inconsistency in DNA repair mechanisms in prokaryotic cells
3. Sporadic presence of detoxifying enzymes in mammalian cell culture systems

4. Variable levels of sensitivity of *in vitro* methods to particular classes of mutagenic agents
5. High sensitivity of *in vitro* tests whose results may not reflect *in vivo* exposures

Other confounding issues of human biology temper the validity of *in vitro* tests for detecting carcinogenic potential. Consider the influence of environmental factors including exposure circumstances, background rates of spontaneous mutations, and the relationship of gene structure and biological function under normal circumstances of existence. These dynamics should enhance the development of regulatory testing schemes to prevent adverse effects associated with exposure to chemicals rather than confuse and inhibit progress toward this goal.

It is interesting that although general principles of mutagenicity risk assessment and guidelines are outlined and published, quantitative risk assessments are rarely made even within regulatory agencies. The volumes of genotoxicity data are currently used to make regulatory decisions based on genotoxic hazard classification or to provide mechanistic information to support quantitative carcinogenicity risk assessments. Significant milestones and progress were achieved with the development of *in vitro* mutagenicity testing technology and with their concordance with *in vivo* models. The future for using these models in human risk assessment portends that speculation is directed in favor of conservative judgments. Consequently, the only guarantee that the controversies, limitations, and agreements among regulatory agencies, industrial concerns, and research institutions will be resolved is the achievement of perfect harmonization. Finally, developing acceptable *in vitro* tests systems to reduce unnecessary animal use is a noble goal, especially if the *in vitro* data for *quantitative* predictions of cell-specific effects are reliable. However, the goal is valid as long as human risk assessment is not compromised.

SUGGESTED READINGS

Ames, B.N., Gurney, E.G., Miller, J.A., and Bartsch, H., Carcinogens as frameshift mutagens: metabolites and derivatives of 2-acetylaminofluorene and other aromatic amine carcinogens, *Proc. Natl. Acad. Sci. USA*, 69, 3128, 1972.

Ames, B.N., Sims, P., and Grover, P.L., Epoxides of carcinogenic polycyclic hydrocarbons are frameshift mutagens, *Science*, 176, 47, 1972.

Ames, B.N., Identifying environmental chemicals causing mutations and cancer, *Science*, 204, 587, 1979.

Ames, B.N., Gold, L.S., and Willett, W.E., The causes and prevention of cancer, *Proc. Natl. Acad. Sci. USA*, 92, 5258, 1995.

Cohen, S.M. and Ellwein, L.B., Cell proliferation in carcinogenesis, *Science*, 249, 10071, 1990.

Weinstein, I.B., Cancer prevention: recent progress, and future opportunities, *Cancer Res.*, 51, 5080s, 1991.

REVIEW ARTICLES

Aardema, M.J. et al., SFTG international collaborative study on *in vitro* micronucleus test III. Using CHO cells, *Mutat. Res.*, 607, 61, 2006.

Bolt, H.M., Foth, H., Hengstler, J.G., and Degen, G.H., Carcinogenicity categorization of chemicals: new aspects to be considered in a European perspective, *Toxicol. Lett.*, 151, 29, 2004.

Brusick, D., Evolution of testing strategies for genetic toxicity, *Mutat. Res.*, 205, 69, 1988.

Carere, A., Stammati, A., and Zucco, F., *In vitro* toxicology methods: impact on regulation from technical and scientific advancements, *Toxicol Lett.*, 127, 153, 2002.

Clare, M.G. et al., SFTG international collaborative study on in vitro micronucleus test II. Using human lymphocytes, *Mutat. Res.*, 607, 37, 2006.

Decordier, I. and Kirsch-Volders, M., The *in vitro* micronucleus test: from past to future, *Mutat. Res.*, 607, 2, 2006.

Farmer, P.B., Committee on Mutagenicity (COM) of Chemicals in Food, Consumer Products and the Environment ILSI/HESI research programme on alternative cancer models: results of Syrian hamster embryo cell transformation assay. *Toxicol. Pathol.*, 30, 536, 2002.

Hartley-Asp, B., Wilkinson, R., Venitt, S., and Harrap, K.R., Studies on the mechanism of action of LS 1727, a nitrosocarbamate of 19-nortestosterone. *Acta Pharmacol. Toxicol.*, 48, 129, 1981.

Kirkland, D.J. et al., Testing strategies in mutagenicity and genetic toxicology: an appraisal of the guidelines of the European Scientific Committee for Cosmetics and Non-Food Products for the evaluation of hair dyes, *Mutat. Res.*, 588, 88, 2005.

Kowalski, L.A., *In vitro* carcinogenicity testing: present and future perspectives in pharmaceutical development, *Curr. Opin. Drug Discov. Devel.*, 4, 29, 2001.

Müller, L. et al., ICH-harmonized guidances on genotoxicity testing of pharmaceuticals: evolution, reasoning and impact, *Mutat. Res.*, 436, 195, 1999.

MacGregor, J.T., Casciano, D., and Muller, L., Strategies and testing methods for identifying mutagenic risks, *Mutat. Res.*, 455, 3, 2000.

Nesnow, S., Complex mixtures of chemical carcinogens, in *Molecular Carcinogenesis and the Molecular Biology of Human Cancer*, Warshawsky, D. and Landolph, J., Eds., CRC Press, Boca Raton, FL, 2006, chap. 14.

Oliver, J. et al., SFTG international collaborative study on *in vitro* micronucleus test V. Using L5178Y cells, *Mutat. Res.*, 607, 125, 2006.

Parry, J.M. and Parry, E.M., The use of the *in vitro* micronucleus assay to detect and assess the aneugenic activity of chemicals, *Mutat. Res.*, 607, 5, 2006.

Preston, R.J., Genetic toxicology, in *Introduction to Biochemical Toxicology*, 3rd ed., Hodgson, R. and Smart, R., Eds., John Wiley & Sons, New York, 2001, chap. 16.

Rothfuss, A., Steger-Hartmann, T., Heinrich, N., and Wichard, J., Computational prediction of the chromosome-damaging potential of chemicals, *Chem. Res. Toxicol.*, 19, 1313, 2006.

Schramke, H. et al., The mouse lymphoma thymidine kinase assay for the assessment and comparison of the mutagenic activity of cigarette mainstream smoke particulate phase, *Toxicology*, 227, 193, 2006.

Wexler, P., The U.S. National Library of Medicine's Toxicology and Environmental Health Information Program, *Toxicology*, 198, 161, 2004.

Williams, G.M., Detection of chemical carcinogens by unscheduled DNA synthesis in rat liver primary cell cultures, *Cancer Res.*, 37, 1845, 1977.

16 Reproductive and Teratogenicity Studies *In Vitro*

16.1 INTRODUCTION

Standard *in vivo* methods for screening substances that are potentially toxic to a fetus (see Chapter 9) are laborious, expensive, and require extensive preliminary planning of experiments to determine the correct endpoint in the reproductive scheme.

Many chemicals cause a variety of pre- and post-natal developmental abnormalities and also interfere with mating and conception. Chemicals that induce structural malformations, physiological dysfunctions, behavioral alterations, and deficiencies in the offspring, at birth or in the immediate post-natal period, are referred to as *teratogens*. Most chemicals are not extensively screened for their teratogenic potential, principally because of the many toxicity tests that are already required. Thus, most pharmaceutical and chemical companies do not routinely screen therapeutic drugs or industrial chemicals, respectively, for potential embryotoxicity during pre-clinical phases. Chemicals not destined to be used clinically are generally not subjected to teratogenic testing.

This means we have no embryotoxic data for the majority of substances used in the marketplace. Consequently, there is significant pressure on regulatory agencies to accept models such as the whole embryo culture described below as screening systems for safety evaluations. A sustained effort in the development of these and other *in vitro* methods into standardized, scientifically validated tools for developmental toxicology is important and crucial. The methodological development of these culture techniques is described in the following sections.

Classical methods for testing teratogenic substances utilize rodents and rabbits, principally because of their relatively short gestational periods (21 to 22 and 32 days, respectively). Depending on the experimental objective, the procedures involve exposure of the female to a potential teratogen either before mating, after conception, during gestation, or in the immediate post-natal period. Alternatively, males are exposed to chemicals when the specific objective is to determine the influence of a chemical on male reproductive capacity. Thus, the start of the exposure sequence depends on the toxicological questions of interest. Also, the trimester during which the chemical has its greatest toxic effects is determined by exposing gravid animals during a specific time of the gestational period. Using this protocol provides answers about the mechanisms of action.

Rodents and rabbits are convenient because of the large sizes of their litters, usually about 8 to 10 embryos in a gravid animal. This allows for scoring of the chemicals according to the number of fetuses affected. For instance, a particular indicator is assessed in each embryo, the minimal accepted value of which is regarded as toxic. The greater the number of embryos that exhibit toxicity by this criteria, the higher the score and the more toxic the chemical. Given the potential number of chemicals that can be screened at the different dosage levels for various indicators in each fetus, the number of experiments can be technically and financially exhausting.

16.2 ALTERNATIVE METHODS FOR EMBRYOTOXICITY TESTING

Alternative methods for embryotoxicity testing and developmental toxicology are divided into three types of systems: (1) whole embryo cultures (WECs), (2) cell cultures, and (3) organ cultures. WEC and organ culture methods enjoy the advantages of representing higher levels of developmental mechanisms and endpoints that adequately reflect organogenesis and morphogenesis. The methods, however, are laborious and rely on the use of animals. In contrast, while primary and continuous cell cultures are relatively easy to perform and depends on minimal or no animal use, they are mechanistically simpler. Thus, lower levels of sensitivity may not be sufficient for detecting many potential teratogens that are more likely to be revealed with WEC methods. Table 16.1 summarizes the advantages and disadvantages of the three systems that are further discussed below.

TABLE 16.1
Advantages and Disadvantages of Alternative Methods for Embryotoxicity Testing

Method	Advantages	Disadvantages
Whole embryo culture (WEC)	Mechanistically more complex; more sensitive to detecting potential teratogens; mimic different gestational periods	Requires significant numbers of animals; laborious, expensive; serum concentration influences results; limited period of embryogenesis in culture
Organ culture	Mechanistically not as complex as WEC; sensitive for detection of potential teratogens; mimics different gestational periods; detects organ specific toxicity	Requires fewer animals than WEC methods; expensive and laborious
Primary and continuous cell culture	Technically easier; requires fewest animals; contributes to understanding mechanistic toxicity	Simpler biological system; less sensitive to embryotoxic effects; devoid of biotransformation capacity

16.2.1   Whole Embryo Culture

Post-implantation embryo culture is a classical method for the detection of potential embryotoxic agents. The technique involves the removal of 1- or 2-week old rodent embryos after which they are acclimated under cell culture conditions for several hours. The viable fetuses are then exposed to different concentrations of chemicals suspected of altering developmental status. The method is dramatic: structural changes occur quickly and teratogenicity is grossly monitored *in vitro* as growth, development, and differentiation progress.

One potential benefit of the method is that it allows for further investigation of substances with teratogenic activities *in vivo*. In addition, culturing embryos *in vitro* provides the opportunity to study the direct actions of teratogens free of maternal neuroendocrine, immunological, and metabolic influences the advantage of which is viewed as a distinct disadvantage outside the realm of cell culture technology. In addition, dose-related growth retardation and frequency of malformations using whole embryo cultures also depend on the periods of gestation of the explanted embryos.

Although the procedure uses growth in culture as a component of the protocol and the methods for growing whole mouse embryos *in vitro* have largely improved in the past two decades, the technique should not be confused with *in vitro* cell culture methods. Embryo cultures still require extensive animal preparation and sacrifice and offer few advantages for large scale prescreening studies. The technology, however, enables embryos to develop remarkably close to what is achieved *in vivo* from gastrulation to early organogenesis stages, while permitting direct chemical and micro-manipulation of embryos. Lastly, procedures involving *pre-implantation* and *early post-implantation* of embryos during organogenesis (described below) may not be adequate for screening chemicals that have teratogenic potentials because they monitor single cell processes that do not reflect the complex nature of embryonic development.

16.2.2   Cell Culture Methods

Because of the problems associated with traditional embryotoxicity testing, efforts have been made to develop *in vitro* cell culture teratogenicity tests. A prerequisite for the use of embryonic tissue in culture is that the isolated organs or cells are maintained in culture for at least 48 to 72 hr. Several methods incorporating isolated cells or organs, e.g., the "micromass" method, are used. The technique is described for the culture of dissociated rodent arm and leg rudiment cells that are capable of synthesizing a variety of extracellular matrix proteins. Protein production is inhibited by most known teratogenic chemicals and measured by classical spectrophotometric analysis or by incorporation of radio-labelled amino acids.

16.2.2.1   Pre-Implantation Techniques

Since the first description of the culture of *pre-implantation* embryos by Brachet, the method has shown its importance in experimental embryology and reproductive biology. The technique is described for the pre-implantation culture of embryos from mice, rabbits, and humans. In general, 3- to 6-day old embryos (4-day morulae and 6-day blastocysts) are removed from superovulated sexually mature animals and transplanted into culture dishes containing media supplemented with bovine serum albumin, serum, or serum-free conditioned medium. The cultures are exposed to the test substance, examined histologically, and monitored for growth and proliferation using biochemical procedures. With the advent of embryonic stem cell technology, however, the procedure may soon lose its popularity.

16.2.2.2 Continuous Cell Cultures

Finite or immortal continuous cells lines are used for monitoring teratogenicity *in vitro*. Several studies now attest to the importance of the potential for these cell lines for screening a variety of chemicals for teratogenic potential. The cell lines include differentiated early human embryonic cells (HFL1, MRC-5, and WI38 fetal human lung fibroblasts) derived from first trimester fetuses and represent the period of gestation when the embryo is most susceptible to developmental anomalies. The fibroblasts are easily maintained in continuous culture, have finite life spans, and are characterized in aging studies. It is important to note, however, that these continuous cell lines may not be suitable cell culture models for screening chemicals with potential toxicity to fetal *development* because the lines are beyond the primary culture stage. Consequently, their responses may reflect basal cytotoxicity rather than a specific effect on growing fetuses.

16.2.2.3 Embryonic Stem (ES) Cells

ES cells are derived from pluripotent cells of early mammalian embryos and are capable of unlimited, undifferentiated proliferation *in vitro*. In chimeras with intact embryos, ES cells contribute to a wide range of adult tissues including germ cells, providing a powerful approach for introducing specific genetic changes into a mouse cell line. Pluripotency allows for the ability of the cells to differentiate to many types and tissues including all three embryonic germ layers. Furthermore, experimental evidence is available for organogenesis of embryonic and adult stem cells to neural, muscle, dermal, epithelial, and bone marrow lineages. Progress on ES cells is prompted, in part, by the understanding that *in vivo* epithelium, epidermal, and mesodermal cells undergo continuous renewal by multipotent stem cells that remain anchored to basement membranes. Thus a single stem cell migrates, differentiates, and commits to all cell line lineages.

The study of embryo stem cells began modestly in the 1960s with manipulation of early rabbit and mouse morulae and blastocysts that adhered readily to plastic tissue culture vessels. They were observed to form sheets of cells that were overgrown by stem cells from the inner cell masses. Whole blastocyst cultures on collagen-coated surfaces produced aggregates of cells that eventually migrated and differentiated into neural, blood, and phagocytic cell types. Not until the inner cell mass was freed and cultured intact did the research community realize that ES cells were capable of possessing good rates of cleavage and could demonstrate morphological and chromosomal stability. In mice, formations of cell aggregates with features of intermediate differentiation are known as embryoid bodies (EBs). Markers for differentiation of pluripotency reveal how neural, cardiac, hematological, and other cell lineages are derived *in vitro*.

The derivation of human ES cells is a relatively recent phenomenon (late 1990s) and these cells have clinical utility because of their rapid colonization and potential for targeting tissues via fetal pathways. Thus, human ES cells offer broad therapeutic potential for clinical applications, the realization of which is a matter of time.

The first description of a stem cell test as an *in vitro* assay for teratogenic potential outlined a method for maintaining mouse ES cells in culture in an undifferentiated or differentiated state. Manipulation of the culture conditions allowed establishment of a population of differentiated cells onto which chemicals could be tested, without the need for continuously establishing primary cultures from mouse embryos. Later, the embryonic stem cell test (EST) was used to detect teratogenic potentials of several compounds in undifferentiated cells. In 2004, ZEBET, in cooperation with the European Centre for Validation of Alternative Methods (ECVAM), coordinated pre-validation and validation studies with three embryotoxicity tests including the EST. The studies concluded that correlation between *in vitro* EST data and *in vivo* data was good and that the EST was applicable for testing a diverse group of chemicals of different embryotoxic potentials.

16.2.3 ORGAN CULTURES

A variety of organ culture models have been developed as reproductive and developmental toxicology test systems. The principles of these models are focused on the improvement of *in vitro* alternative methods that more closely represent the complex process of embryogenesis *in vivo*. The methods require primary animal tissue, however, and have specific technical features that render them more laborious and more difficult to standardize among laboratories. Tests in this group include organotypic models for limb buds, digits, palatal shelves, lungs, intestines, and male and female reproductive organs.

Fetal organs have been maintained in culture in the primary cell stage by explanting them and partially or totally submerging their tissues in growth medium. A tissue explant is placed on a micron filter or in tissue culture plastic vessels coated with extracellular matrix components. This approach is useful in the culture of limb buds from several species of animals in serum-supplemented or chemically defined medium. Morphological development, differentiation, and biochemical parameters are monitored. The variety of primary organotypic cultures isolated successfully include:

1. Cultures of mouse kidneys for studies of renal development and suspected nephrotoxic agents
2. Limb buds for detecting chemical developmental anomalies
3. Gastrointestinal structures for potential teratogens for stomach and intestinal development
4. Cultures of testes and ovaries for reproductive studies
5. Isolation of type II epithelial cells from fetal rat lungs

Rat lung epithelial cells represent specialized cells capable of synthesizing and storing a surfactant composed of phospholipids; its secretion is destined for the pulmonary alveolar space. *In vivo,* the surfactant coats the alveolar walls and decreases the surface tension present at the air–cell interface and thus allows oxygen to penetrate the thin alveolar structures. Consequently, the cells are used in toxicity testing to monitor embryotoxic effects in the lungs, particularly in relation to surfactant production and secretion. Thus, as with other organ-specific primary cultures, their advantage lies in their ability to act as models for detecting target organ toxicities.

Inter-laboratory validation studies involving alternative mammalian organotypic models provide the toxicological community with a variety of methods for inclusion in a battery of tests for testing chemicals with embryotoxic potential.

16.3 VALIDATION OF ALTERNATIVE METHODS FOR REPRODUCTIVE AND TERATOGENICITY STUDIES

The validation of alternative test systems relies on a comparison of sensitivity, specificity, and concordance with documented *in vivo* studies. Since only those chemicals for which sufficient *in vivo* data are available are used for comparisons of developmental toxicity, attempts at *in vitro* validation are limited by the number of test chemicals available. In addition, the most relevant compounds for hazard assessment are those whose developmental toxicities are within marginal toxic ranges and clearly not at the extreme toxic or non-toxic ends. Thus, the ability of *in vitro* screening methods for predicting potential teratogens is hampered not by the development of the testing method, but by the lack of adequate comparative *in vivo* information.

In order to standardize methodologies and define endpoints for *in vitro* analysis, ECVAM defined a prediction model approach that attempts to extrapolate *in vitro* data to *in vivo* studies. The prediction model contains quantitative endpoints that are expected to predict *in vivo* responses. The project outcomes evaluate the performance of a test and are modified when the database of chemicals tested in the *in vitro* system is sufficient, especially when performed with unknown classes of chemicals.

Currently, the most elaborate validation study of embryotoxicity assays performed was financed by ECVAM and includes the EST, the limb bud micromass (MM), and the WEC technique. Each test involved 20 chemical compounds tested in four independent laboratories in a double blind design. Preliminary data indicate that the WEC method displayed the best concordance of *in vivo* classification and *in vitro* test results with 80% correct classifications versus 78 and 71% for EST and MM, respectively. Moreover, strong embryotoxic compounds showed a predictivity of 100% in each of the test systems.

16.4 SUMMARY

The wealth of potential alternative tests available for developmental toxicology testing is in part a reflection of the variety of biological mechanisms involved in embryonic and fetal development. Since it is widely accepted that embryogenesis is a remarkably complex physiological process, it is reasonable to conclude that, with our current biotechnology, some time is required before a single alternative test can replace whole animal testing. Hence, it is more likely, at least in the immediate future, that *combinations* or *batteries* of tests will be able to address the complete range of mechanisms of embryonic and reproductive development.

Because of the unique features of this process, alternative testing in developmental toxicology differs significantly from the principles of *in vitro* dermal and ocular toxicology testing areas that have already enjoyed substantial success in delivering alternative tests for regulatory acceptance. Scientific progress, however, is not ostensibly gradual but is more compelling and abrupt, so that realization of advanced alternative models for reproductive and embryo toxicology may appear on an unforeseeable, yet closer, horizon.

SUGGESTED READINGS

Bremer, S. and Hartung T., The use of embryonic stem cells for regulatory developmental toxicity testing in vitro: the current status of test development, *Curr. Pharm. Des.*, 10, 2733, 2004.

Brown, N.A., Selection of test chemicals for the ECVAM international validation study on *in vitro* embryotoxicity tests, *Altern. Lab. Animals*, 30, 177, 2002.

Flint, O.P., *In vitro* tests for teratogens: desirable endpoints, test batteries and current status of the micromass teratogen test, *Reprod. Toxicol.*, 7, 103, 1993.

Genbacev, O., White, T.E., Gavin, C.E., and Miller, R.K., Human trophoblast cultures: models for implantation and peri-implantation toxicology, *Reprod. Toxicol.*, 7, 75, 1993.

Genschow, E. et al., The ECVAM international validation study on in vitro embryotoxicity tests: results of the definitive phase and evaluation of prediction models, *Altern. Lab. Animal*, 30, 151, 2002.

Genschow, E. et al., Validation of the embryonic stem cell test in the international ECVAM validation study on three *in vitro* embryotoxicity tests, *Altern. Lab. Animals I*, 32, 209, 2004.

Kirkland, D.J. et al., Testing strategies in mutagenicity and genetic toxicology: an appraisal of the guidelines of the European Scientific Committee for Cosmetics and Non-Food Products for the evaluation of hair dyes, *Mutat. Res.*, 588, 88, 2005.

Palmer, A.K., Introduction to (pre)screening methods, *Reprod. Toxicol.*, 7, 95, 1993.

REVIEW ARTICLES

Barile. F.A., Ripley-Rouzier, C., Siddiqi, Z.E., and Bienkowski R.S., Effects of prostaglandin E1 on collagen production and degradation in human fetal lung fibroblasts, *Arch. Biochem. Biophys.*, 265, 441, 1988.

Barile, F.A., Siddiqi, Z.E., Ripley-Rouzier, C., and Bienkowski, R.S., Effects of puromycin and hydroxynorvaline on net production and intracellular degradation of collagen in human fetal lung fibroblasts, *Arch. Biochem. Biophys.*, 270, 294, 1989.

Barile, F.A., Arjun, S., and Hopkinson D., *In vitro* cytotoxicity testing: biological and statistical significance, *Toxicol. in Vitro*, 7, 111, 1993.

Bechter, R., The validation and use of *in vitro* teratogenicity tests, *Arch. Toxicol. Suppl.*, 17, 170, 1995.

Brachet, A., Recherches sur le determinism hereditaire de l'oeuf des mammiferes: development *in vitro* de jeunes vesicules blastodermique du lapin, *Arch. Biol.*, 28, 447, 1913.

Brown, N.A., Teratogenicity testing *in vitro*: status of validation studies, *Arch. Toxicol.*, 11, 105, 1987.

Buesen, R. et al., Trends in improving the embryonic stem cell test (EST): an overview, *ALTEX*, 21, 15, 2004.

Edwards, R.G., Stem cells today: Origin and potential of embryo stem cells, *Reprod. Biomed. Online*, 8, 275, 2004.

Evans, M. and Kaufman, M., Establishment in culture of pluripotential cells from mouse embryos, *Nature*, 92, 154, 1981.

Fein, A. et al., Peri-implantation mouse embryos: an *in vitro* assay for assessing serum-associated embryotoxicity in women with reproductive disorders, *Reprod. Toxicol.*, 12, 155, 1998.

Flick, B. and Klug, S., Whole embryo culture: an important tool in developmental toxicology today, *Curr. Pharm. Des.*, 12, 1467, 2006.

Flint, O. and Orton, T.C., An *in vitro* assay for teratogens with cultures of rat embryo midbrain and limb bud cells, *Toxicol. Appl. Pharmacol.*, 76, 383, 1984.

Newall, D.R. and Beedles, K., The stem cell test: a novel *in vitro* assay for teratogenic potential, *Toxicol. in Vitro*, 8, 697, 1994.

Pellizzer, C., Bremer, S., and Hartung, T., Developmental toxicity testing from animal toward embryonic stem cells, *ALTEX*, 22, 47, 2005.

Piersma, A.H., Validation of alternative methods for developmental toxicity testing, *Toxicol. Lett.*, 149, 147, 2004.

Riecke, K. and Stahlmann, R., Test systems to identify reproductive toxicants, *Andrologia*, 32, 209, 2000.

Rohwedel, J., Guan, K., Hegert, C., and Wobus, A.M., Embryonic stem cells as an *in vitro* model for mutagenicity, cytotoxicity and embryotoxicity studies: present state and future prospects, *Toxicol. in Vitro*, 15, 741, 2001.

Spielmann, H. and Liebsch, M., Validation successes: chemicals, *Altern. Lab. Animals*, 30, 33, 2002.

Tam, P.P., Post-implantation mouse development: whole embryo culture and micro-manipulation, *Int. J. Develop. Biol.*, 42, 895, 1998.

Webster, W.S., Brown-Woodman, P.D., and Ritchie, H.E., A review of the contribution of whole embryo culture to the determination of hazard and risk in teratogenicity testing, *Int. J. Dev. Biol.*, 41, 329, 1997.

17 High Throughput Screening and Microarray Analysis

17.1 HIGH THROUGHPUT SCREENING IN PRECLINICAL DRUG TESTING

17.1.1 INTRODUCTION

High throughput screening (HTS) incorporates a combination of modern robotics, data processing and control software, liquid handling devices, and sensitive detectors that allow efficient and smooth processing of millions of biochemical, genetic, or pharmacological tests in a short time. It provides a rapid, cost-effective method to screen potential sources for novel molecules often used in drug development. Unlike conventional or manual screening procedures, HTS allows for the handling of a wide variety of input samples; each sample is stamped with its own set of data and tracking issues and produces new experimental results. HTS makes it possible to collect large amounts of experimental data related to observations about biological reactions to exposure to various chemical compounds.

In the competitive toxicology and pharmaceutical industries, the time from drug development and preclinical testing to clinical trial is critical. Traditional industrial research approaches do not fulfill the demands of current drug development schedules. HTS uses laboratory automation to collect vast amounts of toxicity and pharmacological data and reliably profiles a range of biological activities and the information eventually is incorporated into large chemical libraries. The availability of these libraries and robotic systems for bioassays allows synthesis and testing of thousands of compounds daily.

17.1.2 PROCEDURES

In preparation for an assay, microwell plates are filled with a biological target such as a protein, intact cells, or an animal embryo. The target absorbs, binds to, or reacts with the compounds in the wells after a predetermined incubation time. Measurements are recorded manually or automatically across all the wells of the plate.* Automated

---

* Manual measurements are often necessary when microscopic examination is used to detect changes or defects in embryonic development caused by the test compounds.

analysis is conducted on several simultaneous samples using indicators such as reflectivity or transmission of polarized light through the target. The data output is recorded as a grid of numeric values, with each number mapping to the value obtained from a single well. High-capacity analysis equipment measures dozens of plates in minutes, thus generating thousands of experimental data points quickly.

17.1.3 EQUIPMENT

The HTS work station consists of liquid handling, plate replication, and plate sealing devices. A variety of microwell plates are used; the 96- to 384-plate wells are the most common. Each well is tagged with experimental targets, chemical compounds, or blanks. In addition, automated liquid handling devices, detection, and other robotic systems are featured. Absorbance plate readers and detectors are used for kinetic measurements in enzyme or cellular assays. Laser scanning fluorescence microplate cytometers offer high content screening at throughputs compatible with primary and secondary screening protocols for fluorescence-based assays.

The robotics features perform autonomous or preprogrammed tasks and are capable of screening combinatorial libraries against screening targets using cell-based assays. Daily average capacities of 480 microplates can generate over 180,000 data points.

17.1.4 APPLICATIONS

Biological and chemical applications for drug discovery and toxicological analysis are applicable for identifying toxicologically or pharmacologically active compounds, antibodies, and genes that modulate particular biomolecular pathways. These results provide starting points for drug designs and for understanding the mechanisms or interactions of target molecules in biochemical processes. In addition, the technology is useful to screen novel drug candidates against libraries of chemicals for their abilities to modify targets (e.g., the ability of a chemical to modify a protein kinase target).

As a method for drug design, biological and physical properties of the targets are screened and a prediction model is constructed to categorize chemicals that interact with the active site (or receptor) of interest. Once a lead molecule series is established with sufficient target potency, selectivity, and favorable drug-like properties, one or two compounds designated "lead" compounds are proposed for drug development; the other compounds are considered "back-ups." It is not unusual for research and development sectors to screen 100,000 to 300,000 compounds per screen to generate 100 to 300 "hits." On average, one or two hits are assigned to the lead compound series. Larger screens of up to 2 million to 3 million compounds [ultra-high throughput screening (UHTS)] are required to generate to 7 to 10 leads. Improvements in lead generation also comes from optimizing library diversity.

17.1.5 FUTURE DIRECTIONS IN HTS

Designing drug candidates for therapeutic interventions and identifying toxic agents have increased the pressure to produce clinical candidates for new drugs and construct models for human risk analysis, respectively. The demand is creating a pressure to identify lead compounds at much faster rates than currently available. This imperative is forcing a technological shift toward higher density plate formats (1536- and 9600-well plates), nanoliter pipetting, and simultaneous imaging.

Assay miniaturization may be capable of facilitating screens of up to 500,000 compounds per week while minimizing reagent costs. Combinatorial chemistry is offering the twin opportunities of developing large compound libraries for screening with faster analogs of lead compounds. Finally, HTS is expanding toward target identification, validation and conversion of assay hits to qualified leads via information generated either within screens or through downstream, high-throughput ADME (absorption, distribution, metabolism, and excretion) and toxicity testing.

17.2 MICROARRAY SYSTEMS

17.2.1 INTRODUCTION

Microarrays are miniature arrays of gene fragments attached to glass slides. The presentation of thousands of gene fragments in a single array allows for the detection of changes in gene expression within a significant fraction of the total genome. The linear arrays of molecules are immobilized at discrete locations on an inert surface that permits simultaneous analysis.

Microarray technology is popular because it is conceptually simple, easy to implement, and is timed well to the growing interest in high throughput parallel studies. In the past decade, microarrays have evolved from membrane-based low density arrays to silicon chip high density oligonucleotide arrays.

17.2.2 TYPES OF ARRAYS

DNA microarrays have DNA molecules immobilized at specific locations on glass or silicon surfaces. They are developed as platforms for studying gene expression and mutation detection and are the subjects of the discussion below. Similarly, *protein microarrays* that incorporate antibodies at immobilized positions have not yet been widely employed.

17.2.3 DNA MICROARRAY DESIGN

DNA microarrays are manufactured into two formats, depending on the type of molecule immobilized. Arrays containing PCR products of 200 to 2000 kilobase pairs are immobilized along the lengths of molecules by conveniently cross linking to the surface of the array (often referred to as a complementary DNA or cDNA array). Alternately, oligonucleotide probes may be synthesized *in situ* on the array or pre-synthesized and then fixed by covalent linkages to the termini. Figure 17.1 outlines a procedure for the production of and protocol associated with a cDNA microarray plate.

The protocol in the cDNA microarray technology begins with fluorescent labeling of mRNA molecules present in the mammalian cell of interest. The labeled mRNA is then pipetted into the wells of the microarray slide and hybridizes to the

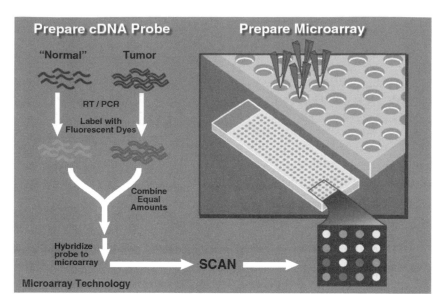

FIGURE 17.1 Production and steps in cDNA microarray technology. (*Source:* National Human Genome Research Institute, U.S. National Institutes of Health, public access at http://www.genome.gov/10000533.)

complementary sequences of the cDNA on the microarray. The intensity of the fluorescence is proportional to the number of mRNA molecules produced by the degree of activity of the gene of interest, indicating hybridization to the corresponding cDNA.

17.2.4 FABRICATION

Microarrays are fabricated using a variety of technologies including photolithography, robotic spotting, and printing (pin-based and piezoelectric pin methods). *Photolithography* is the most widely used method for *in situ* DNA synthesis on arrays to form oligonucleotide microarrays. The technique enables the synthesis of large numbers of high density arrays. *Robotic spotting* uses a robotic arm to deposit presynthesized oligonucleotides and PCR-amplified cDNA fragments on the surface of the array.

The device consists of a stage for holding the printed slides and a robotic arm that uses specialized pins to pick up solutions from a source, usually a 384-well plate. In the *pin-based printing* method, an example of two-system contact printing, steel pins are used to transfer fixed volumes of solution as spots onto the array surface. The flat tips of the pins form spots of a diameter corresponding to the diameter of the tip. The amount of solution absorbed by the pin tip is proportional to the surface tension of the liquid. Subsequently, when the pin is tapped against the array surface, fixed volumes of the solution are dispensed. The volume of the dispensed liquid depends on many factors, most notably on the surface chemistry of the slide and the contact time between the two surfaces. *Piezoelectric pin printing* is an example of a non-contact method that consists of a reservoir for the printing solution and a capillary dispenser. Piezoelectric crystals are apposed closely with the capillary. When a voltage is applied to the crystals, they deform, squeezing together the walls of the capillary and forcing liquid from the orifice.

17.2.5 DETECTION TECHNOLOGY

Most of the current systems are based on fluorescence detection. The schemes are sensitive for biologically active transcripts, safer to handle than radionuclide based methods, and allow multiple sample detections. The two main types of detection systems available for microarrays are scanners and less widely used imagers. Scanners use moving optics or substrates to collect signals from small areas on a slide to compile an entire image.

Alternatively, imagers capture the output from the entire array simultaneously. As with other fluorescent detectors used frequently in cytotoxicology assays, the instrumentation is programmed to account for controls, blanks, normalization of data (equalization of data in comparative experiments), and autofluorescence. The generated data are usually analyzed as scatter plots that represent the comparison of gene expression profiles from different experimental conditions. Typically, fluorescent intensities from different dyes corresponding to different wavelengths are plotted against each other; red color is used to depict over-expression and green is used for under-expression. Figure 17.2 shows a typical cDNA microarray fluorescent glass slide with detection of fluorescent data output.

17.2.6 APPLICATIONS OF cDNA MICROARRAY TECHNOLOGY

17.2.6.1 Gene Expression

cDNA microarrays are most widely used for studying gene expression. The mRNA is isolated from cells of interest in the absence or presence of a toxic agent or drug. The cDNA is prepared from both samples labeled using different fluorescent dyes (most commonly fluorescein green and rhodamine red). As described above, after hybridization of the cDNA molecules to the probe on the array, the ratio of red-to-green fluorescence is proportional to the relative levels of the gene in the test and reference samples. Gene expression in the traditional protocols has many research and clinical applications, some of which are described below.

17.2.6.2 Genotype/SNP Detection

Microarrays are also used for developing chips for detection and screening of single nucleotide polymorphisms (SNPs). A SNP is a DNA sequence variation occurring when a single nucleotide in a genome differs among members of a species or among paired chromosomes in an individual. The technique is applicable for the detection of a single nucleotide difference in sequenced DNA fragments in two alleles from different individuals (e.g., AAGCCTA to AAGCTTA). The differences in hybridization due to single base pair changes are detected using oligonucleotide arrays, provided conditions of hybridization are rigidly controlled. Alternatively, primer exten-

FIGURE 17.2 Example of a microarray printed on the standard 1" × 3" glass slide format; the slides are accessible to standard 1" × 3" microarray scanners. (Courtesy of Agilent Technologies, www.agilent.com.)

sion-based methods that exploit the inabilities of many DNA polymerases to extend a strand where the 3′ terminal nucleotide is not perfectly matched are also used.

17.2.6.3 Detection of Environmental Agents

Environmental toxicological applications of DNA microarray technology have increased due to the ability of the system to monitor for contaminants. For instance, microarrays have found applications for the rapid detection of microbes in wastewater, for indentifcation of pathogens in processed foods, and for monitoring adulteration of food and drug products.

SUGGESTED READINGS

Chan, V.S. and Theilade, M.D., The use of toxicogenomic data in risk assessment: a regulatory perspective, *Clin. Toxicol.*, 43, 121, 2005.

Chen, T., DNA microarrays: an armory for combating infectious diseases in the new century, *Infect. Disord. Drug Targets*, 6, 263, 2006.

Cummins, J.M. and Velculescu, V.E., Implications of micro-RNA profiling for cancer diagnosis, *Oncogene*, 25, 6220, 2006.

Khaitovich, P., Enard, W., Lachmann, M., and Paabo, S., Evolution of primate gene expression, *Nat. Rev. Genet.*, 7, 693, 2006.

Koczan, D. and Thiesen, H.J., Survey of microarray technologies suitable to elucidate transcriptional networks as exemplified by studying KRAB zinc finger gene families, *Proteomics*, 6, 4704, 2006.

Leighton, J.K. et al., Workgroup report: review of genomics data based on experience with mock submissions: view of the CDER Pharmacology Toxicology Nonclinical Pharmacogenomics Subcommittee, *Environ. Health Perspect.*, 114, 573, 2006.

Lettieri, T., Recent applications of DNA microarray technology to toxicology and ecotoxicology, *Environ. Health Perspect.*, 114, 4, 2006.

Modlich, O., Prisack, H.B., and Bojar, H., Breast cancer expression profiling: the impact of microarray testing on clinical decision making, *Exp. Opin. Pharmacother.*, 7, 2069, 2006.

REVIEW ARTICLES

Atterwill, C.K. and Wing, M.G., *In vitro* preclinical lead optimisation technologies (PLOTs) in pharmaceutical development, *Toxicol. Lett.*, 127, 143, 2002.

Breitling, R., Biological microarray interpretation: the rules of engagement, *Biochim. Biophys. Acta*, 1759, 319, 2006.

Bugelski, P.J., Gene expression profiling for pharmaceutical toxicology screening, *Curr. Opin. Drug Discov. Devel.*, 5, 79, 2002.

Coppola, G. and Geschwind, D.H., Microarrays and the microscope: balancing throughput with resolution, *J. Physiol.*, 575, 353, 2006.

Ekins, S., Systems: ADME/tox resources and network approaches, *J. Pharmacol. Toxicol. Methods*, 53, 38, 2006.

Garaizar, J., Rementeria, A., and Porwollik, S., DNA microarray technology: a new tool for the epidemiological typing of bacterial pathogens? *FEMS Immunol. Med. Microbiol.*, 47, 178, 2006.

Gibson, N.J., Application of oligonucleotide arrays to high-content genetic analysis, *Exp. Rev. Mol. Diagn.*, 6, 451, 2006.

Guillouzo, A., Applications of biotechnology to pharmacology and toxicology, *Cell Mol. Biol.*, 47, 1301, 2001.

Irwin, R.D. et al., Application of toxicogenomics to toxicology: basic concepts in the analysis of microarray data, *Toxicol. Pathol.*, 32, 72, 2004.

Jafari, P. and Azuaje, F., An assessment of recently published gene expression data analyses: reporting experimental design and statistical factors, *BMC Med. Inform. Decis. Mak.*, 6, 27, 2006.

Jubb, A.M. et al., Quantitative in situ hybridization of tissue microarrays, *Methods Mol. Biol.*, 326, 255, 2006.

Li, N., Tourovskaia, A., and Folch, A., Biology on a chip: microfabrication for studying the behavior of cultured cells, *Crit. Rev. Biomed. Eng.*, 31, 423, 2003.

Low, Y.L., Wedren, S., and Liu, J., High-throughput genomic technology in research and clinical management of breast cancer: evolving landscape of genetic epidemiological studies, *Breast Cancer Res.*, 8, 209, 2006.

Mans, J.J., Lamont, R.J., and Handfield, M., Microarray analysis of human epithelial cell responses to bacterial interaction, *Infect. Disord. Drug Targets,* 6, 299, 2006.

Maurer, H.H., Screening procedures for simultaneous detection of several drug classes used for high throughput toxicological analyses and doping control: a review, *Comb. Chem. High Throughput Screen.,* 3, 467, 2000.

Meador, V., Jordan, W., and Zimmermann, J., Increasing throughput in lead optimization in *in vivo* toxicity screens, *Curr. Opin. Drug Discov. Devel.*, 5, 72, 2002.

Merrick, B.A. and Bruno, M.E., Genomic and proteomic profiling for biomarkers and signature profiles of toxicity, *Curr. Opin. Mol Ther.*, 6, 600, 2004.

Nielsen, T.O., Microarray analysis of sarcomas, *Adv. Anat. Pathol.* 13, 166, 2006.

Petrik, J. Diagnostic applications of microarrays, *Transfus. Med.*, 16, 233, 2006.

Seo, D., Ginsburg, G.S., and Goldschmidt-Clermont, P.J., Gene expression analysis of cardiovascular diseases: novel insights into biology and clinical applications, *J. Am. Coll. Cardiol.*, 48, 227, 2006.

Shioda, T., Application of DNA microarray to toxicological research, *J. Environ. Pathol. Toxicol. Oncol.*, 23, 13, 2004.

Shockcor, J.P. and Holmes, E., Metabonomic applications in toxicity screening and disease diagnosis, *Curr. Top. Med. Chem.*, 2, 35, 2002.

Sievertzon, M., Nilsson, P., and Lundeberg, J., Improving reliability and performance of DNA microarrays, *Exp. Rev. Mol. Diagn.*, 6, 481, 2006.

Sobek, J. et al., Microarray technology as a universal tool for high-throughput analysis of biological systems, *Comb. Chem. High Throughput Screen.,* 9, 365, 2006.

Walmsley, R.M., Genotoxicity screening: the slow march to the future, *Exp. Opin. Drug Metab. Toxicol.*, 1, 261, 2005.

18 Experimental Design and Statistics

18.1 ESTABLISHING EXPERIMENTAL PLAN

Table 18.1 summarizes several important considerations that are fundamental in planning an *in vitro* experiment. Some of these steps are unique to the development of alternative testing models, while others are simply parts of an adequate plan for conducting any toxicology study.

18.1.1 MATERIALS AND SUPPLIES

The choices of the materials and supplies are important in the preliminary preparations for a study. In fact, the particular manufacturer or distributor of the materials and chemicals is probably not as important as obtaining them from the same supplier. For instance, the manufacturer of cell culture supplies should be reputable, easily accessible, and referenced. In addition, the company should provide the researcher with details of the methods used to prepare, package, store, and ship supplies. The materials should be accompanied by material safety data sheets (MSDSs) that list important packaging and handling precautions for shipping, receiving, and storage. It is also important that a research laboratory maintain records of dates and storage of chemicals. It is also beneficial for consistency of a study protocol that one company provides the majority of the chemicals tested. Among the materials obtained and stocked routinely in sufficient quantities are cell culture supplies, stock cell cultures, chemicals, isotopes, and materials for scintillation counting. Additionally, animal farms can provide an investigator with information about the breeding and housing of animals as well as physical examinations of the animals and a declaration of whether they are certified as pathogen-free.

18.1.2 METHODS

During manipulation of cells in culture, certain details are established and followed prudently and uniformly. The number of cells inoculated into culture vessels depends on the rate of inoculation and the sizes of the flasks or wells. For instance, T-10, -25, -75, and -150 flasks, corresponding to 10-, 25-, 75-, and 150-cm^2 surface area, are suitable for individual experiments or for maintaining cells in continuous passage. Also 6-, 12-, 24-, 48-, 96-, and 384-well plates, corresponding to decreasing surface area per well, are routinely employed in screening assays.

TABLE 18.1
Steps in Planning and Preparing *In Vitro* Experiments

Objective	Considerations
Procuring materials	Standard maintenance of readily available cell culture supplies
Ordering chemicals	Anticipation of rate-limiting chemicals that should be always ready for use
Planning methods	Anticipation of technical, scientific, and administrative scheduling critical for smooth conduct of study
Time constraints	Consideration of time required for initiating and maintaining cultures

In general, cells are usually seeded at 10^4 cells per cm². Contact-inhibited cultures are grown to stationary phase (confluence), unless cell proliferation experiments are performed, and subcultured in appropriate complete medium supplemented with serum and antibiotics if needed. Alternatively, serum-free medium is used in a consistent manner. Depending on the doubling rate of the cell line, the time required for confluent monolayers to grow can be estimated with considerable accuracy. For example, when seeded at one third the confluent density or according to the formula above, most continuous cell lines with appreciable doubling rates reach the stationary phase within 4 to 8 days.

Monolayers of cells are then incubated with increasing concentrations of each chemical in at least four doses (three to six wells per dose) for a predetermined period at 37°C in a gaseous atmosphere defined by the requirements of the medium. The atmosphere may consist of 5 to 20% CO_2 in air unless otherwise stated. A stock solution of the test chemical is prepared in the incubating medium and serial dilutions are made from the stock solution. This method of formulating a stock solution decreases the chances of introducing a repetitive error when adding chemical to each experimental group.

For experiments requiring longer incubation times, the media may be initially sterilized so that aseptic conditions are easily maintained. This is especially important during incubation times of 72 hr or longer. A soluble solution of the chemical is sterilized using micron filters, while insoluble solids are first exposed to ultraviolet light for several minutes in an open test tube or on weighing paper before the addition of buffer.

In time-response experiments, the concentrations for all groups remain the same and the groups are terminated at preset incremental time periods. Before the experiment is terminated, indicators, dyes, fixatives, and reactive substances are added as needed and allowed to incubate with the cultures. The cells are then processed and the reaction product is quantified according to the protocol.

For isotope or fluorescent labeling studies, a radioactive or fluorescent tagged (labeled) precursor substrate is added to the cultures, usually 1 hr after the change from a growth medium to a labeling medium. The precursor may be an amino acid for protein synthesis, a monosaccharide for carbohydrate metabolism, a precursor for lipid biosynthesis, an enzyme substrate, or any molecule designed to trace an intracellular biosynthetic pathway. The labeling medium is a simpler version of the growth medium. Care is taken to avoid the presence of the precursor or parent molecule in the formula. In addition, the serum, generally used during the growth stage is dialyzed to remove trace components that interfere with uptake of the precursor. Other supplements to the incubation medium include buffers, antibiotics, and cofactors necessary for the proper synthesis of the indicator molecules. Cell viability is usually measured in separate cultures and is expressed as a percentage of the control group results. Additional wells without cells are processed with the corresponding medium in parallel as reference blanks.

Several automated or manual cell-based assays are available as prepackaged kits for the spectrophotometric, fluorescent, or luminescent detection of macromolecules representative of cellular biochemical processes. The use of cell-based assays in toxicology and in drug discovery research is important for revealing the mechanisms of toxicological action, development of alternative *in vitro* assays for toxicology testing, and disease mechanisms. Cell-based assays also are important in secondary screening studies of promising drug compounds in the course of drug development. The kits offer plate formats and membranes that allow a wide range of functional cell-based assay protocols to be performed, including:

1. Black multi-screen-fluorescent or luminescent plates with protocols ideal for cell viability and proliferation assays using fluorescent detection or luminescent techniques
2. Migration, invasion, and chemotaxis assay plates suitable for high throughput multiwell filter formats

Some applications of cell-based assay kits are incorporated for studies involving cell viability, proliferation, cell migration, invasion, chemotaxis, whole cell incubation, and whole organism assays. The systems use conventional fluorescent or time-resolved fluorescent detection systems, as well as absorption-based protocols.

More recent fluorescent and luminescent enzyme assays monitor intracellular enzyme activities. These systems are designed to determine cytotoxicity using a variety of fluorescent or luminescent probes including assays for the detection of intracellular enzymes such as calcein-AM, rhodamine-based fluorescent dyes (Molecular Probes, Eugene, Oregon) CellTiter-Glo™, Caspase-3/7, and -8/9 assays (Promega, Madison, Wisconsin). Table 18.2 describes the functional principles of cell-based assays and their value in toxicology testing.

IC_{50} levels and concentration effect curves for each cell line are determined with the probes available in the kits. Results are coordinated by monitoring several indicators that also contribute to an understanding of the mechanisms of toxicity. This information in combination with inter-laboratory comparisons allows for the comparison of the reliability of cell-based assays as indicators for traditional *in vitro* cytotoxicity testing methods. Thus the convenient and rapid performance of a test is critical for incorporation into a battery of tests for screening acute human chemical toxicity.

18.2 EXPERIMENTAL DESIGN

18.2.1 MEDIUM AND CHEMICAL COMPATIBILITY

Most technical problems related to experimental design and execution of toxicology testing experiments involve the solubility or miscibility of the chemicals with the

TABLE 18.2
Cell-Based Assays and Their Utility in Toxicology Testing

Assay	Description
Calcein-AM (Molecular Probes Inc.)	Soluble dye, freely enters intact cell membranes and fluoresces upon cleavage with intracellular esterases; reliable viability indicator
SYTOX™ (Molecular Probes Inc.)	High affinity nucleic acid fluorescent stain; penetrates compromised cell membranes but will not cross membranes of live cells; valuable dead-cell green fluorescent indicator when used simultaneously with other measures of cytotoxicity
Rhodamine-110 series of dyes (Molecular Probes Inc.)	Indicators for intracellular proteinases; upon cleavage with trypsin, plasmin/cathepsin, or other intracellular enzymes, non-fluorescent bisamide derivatives of dyes are converted to fluorescent molecules, with corresponding increases in fluorescence emission
Caspase-3/7 and Caspase–Glo 8/9 (Promega Comp.)	Capable of detecting and differentiating cells undergoing apoptosis or necrosis in the presence of toxic compounds; assay based on cleavage of pro-fluorescent peptide–rhodamine 110 substrate by caspase-3/7, resulting in green fluorescent signal and cleavage of proluminogenic caspase 8/9, resulting in luminescence
CellTiter-Glo™ (Promega, luminescent cell viability)	Homogeneous bioluminescent protocol designed for measurement of ATP; involves addition of luciferin substrate and stable luciferase directly to cells; bioluminescent glow is directly proportional to amount of ATP released from viable cells

incubation medium. Solid chemicals that are insoluble in water, such as paracetamol (acetaminophen) and acetylsalicylic acid (aspirin) present with dissolution problems, especially at higher concentration levels. In order to improve the solubility of these chemicals without the addition of a solvent that may potentially interfere with interpretation of the final results, a variety of techniques are employed, as described in Table 18.3.

18.2.1.1 Micronization

This procedure increases the surface area of a solid and can improve the solubility of a powder at low doses. A solution of the chemical is prepared in ethanol or dimethlysulfoxide (DMSO) for each group and the solution is evaporated to dryness under a stream of nitrogen gas. The remaining lyophilized powder is then dissolved more readily in medium using constant stirring at 37°C for at least 1 hr prior to incubation.

18.2.1.2 Solvents

Occasionally it may be necessary to dissolve a test chemical in a solvent (such as ethanol, isopropanol, and DMSO) that is miscible with the incubation medium. A stock solution of the substance is prepared in the solvent and appropriate aliquots are then distributed to the corresponding experimental groups. It is important, how-

TABLE 18.3
Technical Manipulations to Improve Solubility and Miscibility of Media and Chemicals

Manipulations	Chemical Formulation	Description
Micronization	Solid dissolved in alcohol or DMSO	Evaporation of stock solution of chemical in alcohol, using constant stirring at 37°C for 1 hr or under N_2 gas stream until powder is lyophilized
Stock solution in ethanol, isopropanol, or DMSO	Solid dissolved in solvent	Concentrated stock solution of test chemical is formulated and aliquots are added to media of test groups as required; solvents are equally distributed among all groups including controls
Sonication	Organic liquid	Quantity of immiscible organic test liquid is added to medium and sonicated using an ultrasonic probe
Mineral oil overlay	Organic liquid	Quantity of volatile or organic test liquid is added to medium and mineral oil is layered over surface; flasks are gassed and sealed individually

Note: DMSO = dimethylsulfoxide.

ever, that the relative effect of the solvent alone is determined in control groups, even at its lowest concentration as part of the incubation medium. Also, all experimental groups are equilibrated with the same amounts of solvent including the controls. This will negate any minor effects of the solvent alone on the cells.

18.2.1.3   Sonication

Certain organic liquids such as xylene and carbon tetrachloride used as test chemicals are immiscible with water and culture media. Miscibility is improved by subjecting a stock solution of the chemical to sonication. This manipulation requires the use of an ultrasonic processor equipped with a longer probe for culture tubes that is inserted into the immiscible liquids for as few as 10 sec at 10 to 50 W of power. The ultrasound energy completely homogenizes the mixture and allows enough time for adequately dispersing an aliquot of the "dissolved" test substance solution into the incubation medium. On occasion, the liquids separate out during the incubation time, in which case the mixture is resuspended and added again to the cultures.

18.2.1.4   Paraffin (Mineral) Oil Overlay

More volatile chemicals such as alcohols, xylene, and chloroform are inconvenient because they permeate the incubator atmosphere and interfere with adjoining wells. The evaporation and thus the concentration of the chemical in the medium can be controlled by incubating the chemicals with cells previously grown in a 25 cm^2 (T-25) flask that is then overlaid with a thin layer of paraffin oil (light mineral oil). The oil suppresses evaporation of volatile molecules at the surface by increasing the surface tension. The flasks, however, must be individually permeated with a stream of CO_2 gas corresponding to the gaseous atmosphere of the incubator and sealed with screw caps. This requires some forethought: it is necessary to anticipate which chemicals are tested and which vessels are used to grow the cells. In addition, a separate gas tank formulated for the incubation medium is required for these studies.

The oil overlay produces enough surface tension to prevent evaporation of volatile chemicals incompletely dissolved in the medium. The partial pressure of the chemical in a liquid, however, determines to what extent the test substance partitions between the air and gas phases. Ultimately this partitioning affects the amount of substance dissolved in the incubating medium and in contact with the cell layer. Essentially, it is necessary to verify the actual concentration of the chemical in the liquid and the air space above the cell layer in a series of control flasks. The method of choice includes *head-space gas chromatography*. Accordingly an accurate estimation of the kinetics of air-liquid distribution is computed.

18.2.2 Calculation of Concentration Range

It is not always necessary to test wide ranges of logarithmic concentrations of chemicals, especially when resources are limited. For instance, the LD_{50} value of a known substance is used to estimate the human equivalent toxic concentrations (HETC) in plasma — the concentrations of a chemical used to initiate *in vitro* toxicology testing experiments are derived from rodent LD_{50} values according to the following formula:

$$HETC = [(LD_{50})/V_d] * 10^{-3}$$

where HETC is the estimated human *equivalent* toxic concentration in plasma (mg/ml); LD_{50} is the 50% lethal dose in rodents (mg/kg, intraperitoneal or oral); V_d is the volume of distribution (L/kg); and 10^{-3} represents the constant for conversion into ml (L/ml).

Thus the calculated HETC value is an estimate of the toxic human blood concentration equivalent to the LD_{50} dose administered to a rodent based on body weight. The HETC is then used as a guideline for establishing the concentration range for each group of an experiment. HETC values are also used as a method for converting animal LD_{50} data into *equivalent* human toxicity information, which is valuable as a procedure for comparing *in vitro* IC_{50} values against human lethal concentrations derived from clinical case studies and poison control centers. Essentially, the formula offers a mechanism for converting *in vivo* rodent data to equivalent human toxicity concentrations that can be employed as a guideline to compare with *in vitro* data in their ability to predict human toxicity.

18.2.3 Determination of Inhibitory Concentrations

Figure 18.1 shows a typical concentration effect curve generated for a representative chemical using an indicator such as cell viability for basal cell functions. The concentration necessary to inhibit 50% (IC_{50}) of the newly synthesized macromolecules is then calculated based on the slope and linearity of the regression line. The

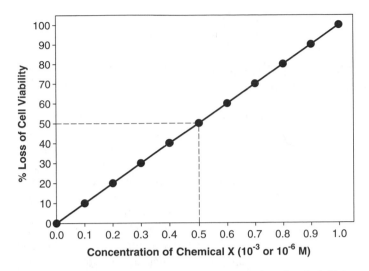

FIGURE 18.1 Concentration effect curve for a representative chemical X in an *in vitro* cytotoxicology experiment using loss of cell viability as indicator. The inhibitory concentration at the 50% level (IC_{50}) is determined from this plot.

symmetry of the line is estimated from the *regression analysis* of the plot: percent loss of cell viability versus concentration.

The values for percent loss of cell viability are derived by converting the absolute value of the measured response to a fraction of the control value, that is, the measured value of the group at the lowest or highest chemical concentration. Using the control values as the 0% or 100% level, respectively, the responses from all subsequent groups are compared to these controls. Thus, absolute values are transformed into relative percentage values. This presents an added advantage such that different experiments are comparable even when the absolute values of the control groups are not identical. When the response is plotted against the concentration used for that group, the regression line is obtained. Similarly, the IC_{50} is estimated by extrapolating from the 50% response.

The plots for some representative substances grouped according to chemical classification in Figure 18.1 show that the regression line is significantly accurate so that the IC^{50} may be computed with statistical confidence.* In a similar manner, the 25 and 75% inhibitory concentrations (IC_{25} and IC_{75}, respectively, not shown) are also calculated to demonstrate the lower and upper cut-off limits. In addition, the cut-off points aid in the determination of the range of toxicity. The greater the difference between the upper and lower inhibitory concentrations, the wider the range of toxicity for a particular chemical. A chemical possessing a smaller difference between the IC_{25} and IC_{75} has a narrower range of toxicity. This implies a narrower margin between non-toxic and toxic doses or between toxicity and lethality.

---

* The calculated correlation coefficient [r value] for this theoretical plot is +1.00.

When the IC_{50} is not bracketed in the concentration range used for the chemical, the experiment is repeated with adjusted sets of concentrations.

18.3 HYPOTHESIS TESTING AND STATISTICAL APPLICATIONS

Many statistical tests are available for analyzing a set of experimental data. The choice of combination of statistical methods used, however, depends on the objectives of the study and the direction of the composition of the dataset. Ostensibly, some forethought is required during the initial experimental planning so that the appropriate statistical approach is employed. The calculations are then incorporated into the study either alone or collectively with other statistical parameters, thus reinforcing the validity of the experimental conclusions.

In addition, the type and number of statistical tests engaged will reduce the chances that a non-significant result is declared significant, especially when individual experiments are repeated. It is an innate consequence, however, that the more often the same statistical calculation is repeated for each experiment, the greater the chances of rejecting the true null hypothesis. Table 18.4 defines some of the fundamental terms used to describe and interpret statistical parameters in toxicology testing.

18.3.1 DEVELOPMENT OF HYPOTHESIS TEST

The purpose of hypothesis testing in cytotoxicity studies is to enable an investigator to reach a decision about a population by examining a sample of that population. The hypothesis, referred to as the *null hypothesis (Ho)* or *hypothesis of no difference,* is constructed as a statement about one or more populations. In the process of the statistical test, the null hypothesis is either rejected or accepted. If the testing procedure leads to rejection, the data are construed as not compatible with the null hypothesis, but are supportive of some other as-yet-undetermined hypothesis. Any experimental treatment protocol in toxicology testing acts as an example of a hypothesis, the easiest of which are the toxicological changes incurred by a cell or organism as a result of the influence of a chemical. In general, the null hypothesis is established for the explicit purpose of being discredited or supported.

The hypothesis is also concerned with the parameters of the selected population. The form of the parameters established as quantifiable data information determines the choice of tests used. In hypothesis testing of univariate parametric tests, the populations are normally distributed and the data are continuous. Each data point has a discrete number that implies a measurable relationship to other numbers among the populations. In hypothesis testing of non-parametric tests, the data are ranked or categorical. These types of data are not continuous, are presented in the form of contingency tables, and are collected and arranged so that they may be classified according to one or more response categories.

18.3.2 PARAMETRIC CALCULATIONS

The selection of parametric and non-parametric tests in toxicology testing studies is based on the initial determination of the availability of continuous or discrete data.

TABLE 18.4
Fundamental Statistics Terms for Interpreting and Understanding *In Vitro* Cytotoxicity Testing Data

Statistical Term	Definition
Independent variable	Characteristic or measurable parameter assigned to different values within test system; also known as *random variable* when values are obtained as result of chance factors; e.g. designated concentration of chemical
Dependent variable	Value resulting from test measurement protocol; relies on value of independent variable; also known as *observation* or *measurement*, e.g., measurement of cell proliferation or enzyme activity after exposure to chemical
Continuous data	Data plotted as discrete points on a line, including any measurable parameter; can assume any of an infinite number of values between two fixed points, e.g., body weight, protein concentration
Discontinuous data	Measurable parameters as with continuous data; have only fixed values, with no possible intermediate values, e.g., number of cells surviving treatment exposure
Probability	Frequency with which a particular value is assigned specified by appropriately sized sample
Frequency distribution	Table or graph in which values of variable are distributed among specified class intervals; set of non-overlapping observations
Hypothesis testing	Determination of whether statement about one or more populations is compatible with available data; hypothesis is concerned with parameters of populations about which statement is made; testing aids investigator in reaching decision about population by examining sample from that population
Null hypothesis (hypothesis of no difference)	Statement about one or more populations supporting contention no significant differences exist between or among groups of sample populations subjected to testing parameters
Significance level	All possible values a test statistic can assume are graphed as points on horizontal axis of distribution and divided into acceptance and rejection regions (see Figure x.x); decision as to which experimental values (observations) are assigned to regions is made according to selected level of significance designated level; level specifies area under curve of distribution of test statistic that is above values on horizontal axis constituting rejection region; the most common significance levels are often set at 0.001, 0.01 or 0.05 corresponding to 99.9, 99.0, and 95% confidence intervals, respectively

Thus, parametric testing focuses on estimating or testing a hypothesis about one or more population parameters. A valid assumption of these tests is that the sampled populations should be approximately normally distributed as in the "entire" population. Another requirement is knowledge of the form of the population providing the samples for the basis of the inferences. Some of the more common parametric tests engaged in statistical analysis are discussed below.

18.3.2.1   Regression Analysis

Regression analysis or linear regression is one of two statistical tests, along with correlation analysis, that yields information about the relationship of two variables.

The variables are usually normally distributed sample populations that take the form of experimental groups. For example, in toxicology testing, the experimental groups include a control group to which no chemical is added and treatment groups, each of which is capable of generating data points. The variables in question consist of measurable responses of cells to a chemical (the Y or *dependent variable*) as a function of the concentration (the X or *independent variable*).

Thus regression analysis facilitates the identification of the probable form of the relationship between these variables, with the ultimate objective of predicting or estimating the value of the dependent variable in response to a known value of the independent variable. As in the case of *in vitro* cytotoxicology experiments, the IC_{50} is estimated based on the regression analysis of the concentration effect curves. Thus, an estimate of the concentration necessary to inhibit 50% of the measured response is calculated within the confidence intervals.

Often, the raw data of concentration effect relationships are not linear, necessitating transformation of the data into logarithmic value scales. Usually, the *logs* of the concentrations are plotted on the X axis against the probability *(probit)* scale, such as percentage response of the control value (Y axis). Such manipulations render the information more amenable to linear interpolation and allow for calculations of estimated IC_{50} values with greater statistical confidence.

18.3.2.2   Correlation Analysis

Correlation analysis and the calculation of the correlation coefficient *(r* value) are computed to determine the strength or significance of a regression analysis. The *r* value dictates the relative degree of correlation between the variables at a predetermined level of significance. The largest absolute value that *r* can assume is ±1.0, resulting when all the variation in the dependent variable is explained by the regression analysis.

In this case, all the experimental observations are reduced to the regression line, within acceptable error limits. The lower limit of *r* is 0, which is obtained when the regression line is horizontal. The closer the *r* value approaches ±1.0 (−1.0 represents an inverse correlation), the greater the probability that the variation of the *Y* values is explained by the regression. This indicates that the regression analysis accounts for a large proportion of the total variability in the observed values for *Y.* Conversely, a small *r* value (close to zero in the positive or negative direction) does not support the regression and is viewed with less confidence. Additional objective statistical tests for significance, however, are the *hypothesis test for* $\beta = 0$, the *analysis of variance,* and the *Student's t-test.*

18.3.2.3   Hypothesis Test for $\beta = 0$

In addition to calculating the correlation coefficient *(r* value) for each concentration effect curve, the hypothesis test for $\beta = 0$ is applied as an alternative test of the null hypothesis of zero slope between two variables, X (log of concentration) and Y (percentage response). The calculation is based on the slope of the sample regression equation. The test statistic, which is normally distributed as with the Student's *t,* is

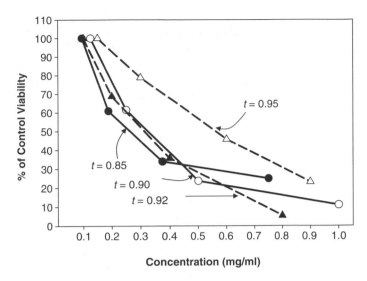

FIGURE 18.2 Graded dose–response curves for different representative chemicals. The percent of viable cells capable of proliferating in a cell culture system decreases in response to increasing concentrations of chemicals.

calculated and compared to minimum values of t at n – two degrees of freedom for the 95 and 99% confidence intervals or greater. Figure 18.2 illustrates concentration effect curves generated for representative chemicals using a cell viability indicator as a measure for cytotoxicity.

The correlation coefficient, the test statistic (t value), and inhibitory concentrations are calculated from the transform regression analysis of each plot (percent of control versus log of concentration). Different sets of concentration levels are repeated for each chemical until the test statistic for the plot is significant at the 95% or 99% confidence interval at this probability. The test statistic is then compared to minimum values for t for the number of pairs of data involved minus two (or n – two degrees of freedom).

If a test statistic is larger than the table value at the predetermined confidence interval, the relationship between the two groups is significant, that is, the effect of the chemical on cell viability is inversely proportional to the change in concentration of the chemical. If the test statistic is smaller than the table value, the two variables have no relationship and the true null hypothesis is not rejected.

18.3.2.4   Analysis of Variance (ANOVA) and *F*-Test

Analysis of variance compares the total variation present in three or more groups of data partitioned into several components. The data are initially evaluated as continuous, independent, and normally distributed. Each component is associated with a specific source of variation so that each source contributes to a portion of the total variation. The magnitude of the contributions is used to calculate the F distribution and compared to a table of F values.

As with the hypothesis test for determination of significance, if the calculated value of F is greater than the critical value for F at a given significance level, then the null hypothesis of equal population variances is rejected. This suggests that the computed F value does not represent a rare event brought about by chance, and that there is a significant difference *among* the groups in the statistical comparison. The Student's t test is then computed to determine *which* groups are significantly different. An F value that is smaller than the critical value for F at a given significance level indicates no significant differences among the groups.

18.3.2.5   Student's *t*-Test

A variety of tests (referred to as *post hoc* tests) are incorporated into experimental procedures to compare the variations of data groups when the F-test calculated from ANOVA is significant. The most popular of these is the Student's t-test or t-distribution. The t-test is used to compare two groups, control and treatment groups or two treatment groups. Two types of Student's t-test are used, depending on the nature of the data points within the groups. If each observation generates corresponding numbers of continuous data and the number of observations is equal, the *paired* test is employed and the degree of freedom is represented as $N - 1$, where N is the number of observations within the groups. When the observations are not equal and the data are continuous but randomly distributed, the unpaired t-test is used. The degrees of freedom are represented as $N_1 + N_2 - 2$, where the number of observations of data between groups is not equal or paired.

As with the calculation of the F value, the value of t is compared to the values in the t-distribution table at the appropriate significance level (α) and degree of freedom. If the calculated t-value is greater than the critical value for t at a given significance level, then the null hypothesis of equal sample variances is rejected, signifying a significant difference between the two groups and the changes on the dependent variable are directly or indirectly proportional to changes of the independent variable. A t-value that is smaller than the critical value for t at a given significance level indicates no differences between the two groups.

18.3.3   Non-Parametric or Goodness-of-Fit Calculations

Non-parametric statistical analysis is exactly analogous to parametric statistical methods but involves the analysis of ranked data. Traditionally known as *goodness-of-fit* tests, these analyses are based on observed distribution of frequencies in treatment groups and their compatibility with a predetermined or hypothesized distribution. The procedures used for reaching a decision to reject or accept the null hypothesis consist of placing the observations into mutually exclusive categories and noting the frequency of observed values in each interval. The frequencies are then compared with available knowledge of normal frequency distributions that would be expected if the samples were obtained from a normal distribution. The discrepancies between what was observed and expected are used to calculate the test statistic that is then compared to minimum values in a standardized table of values for that particular test at a pre-assigned significance level.

18.3.3.1 Chi-Square (χ^2)

The chi-square test for independence of classification is a widely used goodness-of-fit test based on a cross tabulation of observed frequencies of two (2×2) or more ($R \times C$) variables. This cross tabulation is also called a contingency table and arranges the discontinuous frequency data in rows (F) of control and treatment groups and in columns (C) of observed and expected frequencies. The degrees of freedom are calculated as $(R - 1) \times (C - 1)$. Chi-square is applied to normal, binomial, and Poisson distributions, depending on the number of categories. When χ^2 is greater than the tabulated value of χ^2 at a predetermined level of significance (α), the null hypothesis is rejected at that level and the groups are considered significantly different with respect to their treatment protocols. When the null hypothesis is true, the discrepancy between observed and expected frequencies is distributed approximately as the calculated χ^2.

Some underlying assumptions for the chi-square distribution include that the data are univariate and categorical and that the groups are approximately the same size with at least 50 total observations. As with other statistical analyses, the data should be collected by random independent sampling methods.

18.3.3.2 Wilcoxon Rank Sum

As with the chi-square, the Wilcoxon rank sum test is used for comparison of groups of data not normally distributed when the data fall within established groups with set ranges of observations. Data for the experimental comparison groups are initially ranked in increasing order and assigned rank values. The sum-of-ranks values are calculated and compared to the two limit table values to determine whether the groups are significantly different at a given significance level (α).

18.3.3.3 Kruskal-Wallis Non-Parametric ANOVA

As with the parametric analysis of variance, the Kruskal-Wallis non-parametric ANOVA is used to reject or accept the null hypothesis that maintains that several group means are equal. The calculation is initially performed on discontinuous categorical groups of data to determine whether significant differences exist among the groups. The test proceeds by arranging the observations of three or more groups in a single series in order of magnitude from smallest to largest. The observations are replaced by ranks starting from the first. The ranks are then assigned to observations in each group and added separately to yield a sum of the ranks.

The test statistic (H) is computed and compared to a table of H values. When the calculated H value is greater than the table value for the number of observations in the study, the null hypothesis is rejected, suggesting a significant difference between the groups. The test is usually followed by another non-parametric test such as a 2×2 *chi-square* or *distribution-free multiple comparison*, depending on the number of groups, to determine which groups exhibit significant differences. The Kruskal-Wallis ANOVA is used for both small and large samples. The test is weakened when many tied ranks occur.

SUGGESTED READINGS

Barile, F.A., Introduction to In Vitro Cytotoxicology: Mechanisms and Methods, CRC Press, Boca Raton, FL, 1994.

Baselt, R.C., Disposition of Toxic Drugs and Chemicals in Man, 7th ed., Biomedical Publications, Foster City, CA, 2004.

Chow, S.C. and Shao, J., Statistics in Drug Research: Methodologies and Recent Developments, CRC Press, Boca Raton, FL, 2002.

Daniel, W.W., Biostatistics: A Foundation for Analysis in the Health Sciences, 8th ed., John Wiley & Sons, New York, 2006.

Gad, S.C., Statistics and Experimental Design for Toxicologists and Pharmacologists, 4th ed., Taylor & Francis, Boca Raton, FL, 2005.

Lewis, R.J., Jr., Hazardous Chemicals Desk Reference, John Wiley & Sons, New York, 2002.

Skillings, J.H. and Weber, D., A First Course in the Design of Experiments: A Linear Models Approach, CRC Press, Boca Raton, FL, 1999.

The Merck Index, 14th ed., O'Neil, M.J. et al., Eds., Merck Research Laboratories, Whitehouse Station, NJ, 2006.

REVIEW ARTICLES

Bernauer, U., Oberemm, A., Madle, S., and Gundert-Remy, U., The use of *in vitro* data in risk assessment, *Basic Clin. Pharmacol. Toxicol.*, 96, 176, 2005.

Elliott, M.R., Joffe, M.M., and Chen, Z., A potential outcomes approach to developmental toxicity analyses, *Biometrics*, 62, 352, 2006.

Hayashi, Y., Designing *in vitro* assay systems for hazard characterization: basic strategies and related technical issues, *Exp. Toxicol. Pathol.*, 57, 227, 2005.

Linkov, I. and Satterstrom, F.K., Weight of evidence: what is the state of the science? *Risk Anal.*, 26, 573, 2006.

Rietjens, I.M. and Alink, G.M., Future of toxicology: low-dose toxicology and risk–benefit analysis, *Chem. Res. Toxicol.*, 19, 977, 2006.

Suter, W., Predictive value of *in vitro* safety studies. *Curr. Opin. Chem. Biol.*, 10, 362, 2006.

van der Hoeven, N., Current issues in statistics and models for ecotoxicological risk assessment. *Acta Biother.*, 52, 201, 2004.

19 Standardization and Validation of Alternative Methods

19.1 INTRODUCTION

19.1.1 DEVELOPMENT OF ALTERNATIVE TEST METHODS

In the past 20 years, the toxicological community has witnessed an unprecedented burgeoning of development of various types of non-genetic, systemic, and local *in vitro* methods as alternatives to animal testing. Although traditionally founded as mechanistically based cell culture procedures, many of them were adapted as empirical measures of many types of toxicity based on the assumption that their targets were similar to those encountered *in vivo*. Some of these protocols are now well established, for example, the use of cell culture models to screen or predict for dermal and ocular local toxicity. Other proposed alternative *in vitro* tests are currently undergoing standardization of the procedures within single laboratories and among institutions, such as testing for acute systemic and target organ toxicity (including nervous system, liver, heart, and kidney) and various toxicokinetic tests. About 10 to 20 test methods have been developed for each system, now totaling about 100 to 200 different quantitatively reliable procedures.

Some of these techniques were rapidly adopted by the pharmaceutical and cosmetic industries as preliminary toxicology screening tools and used in parallel with conventional animal testing protocols. Based on the successful comparisons of the data with traditional *in vivo* procedures, the newly developed alternative tests are accepted or acknowledged by European Union and United States regulatory agencies for formal validation proposals. It is important to remember, however, that many of the same methods have been in practical use for prediction of acute toxicity, eye irritancy, phototoxicity, and teratogenicity for several years based on their abilities to screen for potentially toxic substances. In fact, the *U.S. Pharmacopoeia* recommends cellular tests for local and systemic toxicity of plastic devices as tests of first choice. Nevertheless these systems represent only a small fraction of the procedures recently made available and cited in the literature and have not been subjected to evaluation by regulatory, industrial, or academic concerns. Unlike the procedures that have been evaluated and have withstood scientific scrutiny, none of the latter methods have gained wide acceptance by the same scientific and governing communities.

19.1.2 STANDARDIZATION OF TEST PROTOCOLS

In general, individual proprietary and academic laboratories develop and propose *in vitro* procedures aimed at specific toxic endpoints. The principle is to analyze a set of chemicals and their effects on cellular components. These evaluations involve testing for injury to basal cell or organ-specific functions, alterations of non-cellular structures, inflammatory reactions to the injuries, and the toxicokinetics of the chemicals in the target tissues. A cell culture system is then constructed to simulate a composite of these factors, or perhaps only one factor, and thus becomes a partial test for the toxicity in question. *In vitro* eye irritancy procedures such as the HET-CAM test were developed in this manner as composite protocols. Other systems such as tests of the corneal epithelium, protein matrix simulating corneal structure, and tests of inflammatory responses were based on the detection of toxicity on focused targets in the eye.

It is lengthy and expensive for individual laboratories to maintain this effort, especially when standardization of the test methods is essential for proposal as a toxicology screening procedure. In the case of eye irritancy, a laboratory must submit for peer review a proposal that the method will be able to screen or predict eye toxicity in humans as well as the existing and accepted Draize procedure. Essentially, a fledgling protocol requires validation, that is, the evaluation of the technical reliability of the method as well as the relevance to its stated purpose. A composite method, for instance, requires testing of hundreds of carefully selected chemicals and groups of chemicals with similar structural classifications or applications for potential to cause human eye irritancy. The results are then compared to those obtained with the Draize method and to existing data for human toxicity. If a test has introduced only one component of the target injury, the other factors of the *in vivo* data are also incorporated into the validation scheme.

For most laboratories, this demanding endeavor is not a realistic goal because the availability of resources is often limited. Most laboratories, therefore, perform preliminary and limited evaluations of their methods as feasibly possible. For instance, 20 to 50 chemicals are tested with the proposed protocol and compared with available animal toxicity data for the same substances, thus presenting a method that has undergone a cursory evaluation. The objective, therefore, is not to propose a scheme for regulatory acceptance, but rather to introduce a blueprint to the scientific and industrial community with the hope that it will be further evaluated for practical and general use as part of a toxicity testing protocol.

The bottleneck to implementation of alternative *in vitro* models into general and local toxicity testing designs is a lack of well defined validation proposals, caused primarily by a paucity of resources of individual laboratories devoted to this purpose. Also, human toxicity data for most types of systemic and local toxicity are scarce, which makes it more difficult to accept screening tests for predicting acute or chronic toxicity. Finally, it is important to note that poorly defined validation processes are impediments to future acceptance of effective *in vitro* testing.

Accordingly, it is not difficult to understand why so many alternative models for *in vitro* testing are not accepted into the general toxicity testing establishment. The explanation lies in a complex set of scientific and regulatory criteria in which newer *in vitro* methods are rigorously and justifiably compared to the theoretical bases that form the fundamentals of all traditional toxicological inquiry while the inscrutable organization of the regulatory agencies inevitably leads to delay. The following discussion presents an explanation of the criteria and components of an alternative *in vitro* validation system, the status of current multilaboratory validation programs, and a proposal for future validation schemes and legislative acceptance.

19.2 VALIDATION PROCESS

19.2.1 Definitions

The *validation* term is derived from the original application used for *in vitro* geno-toxicity and mutagenicity tests. It was introduced into the non-genotoxic arena to denote relevance and reliability of a test for a specific purpose. *Reliability* refers to the reproducibility of a test within the same laboratory and between laboratories: intra- and inter-laboratory variation. *Evaluation* is a process whose aim is acceptance of the test by the scientific community, although the term does not necessarily imply achievement of this goal.

Although validation suggests acceptance by the scientific community is an absolute goal as defined by regulatory agencies, the term is used most often by investigators as a process of appraising *in vitro* methods without the demand that the process be complete to reach this goal. Certain references recently recommended that the term should refer to a *process* used to evaluate the reliability and relevance of *in vitro* tests. This course of action includes the necessary investigatory steps aimed at acceptance of the method in the general scientific and industrial communities rather than an absolute attainable objective. For clarification, Table 19.1 defines some important terms used to explain the validation process and function in the development of alternative test methods.

19.2.2 Vital Components of Validation Program

19.2.2.1 Relevance and Reliability

The importance of the relationship of relevance and reliability is necessarily linked to the concept of validation. For instance, a test that demonstrates favorable results for a large number of chemicals or chemical groups implies considerable reliability. In contrast, a very reliable method may not be useful if it has not demonstrated some relevance to the *in vivo* situation. Many recently developed cytotoxicity testing protocols with objective endpoint determinations from established laboratories are reproducible and their results do not vary significantly among laboratories.

Closer inspection of the protocol, however, reveals that some of these methods lack relevance. Therefore, while relevance and reliability are equally important in the validation process, the concept of relevance is a rate limiting factor that allows little tolerance for deviation from the goal of validation. Because of this criterion, most laboratory validation studies arising from individual laboratories evaluated only relevance and were presented for scientific review with good reasons to believe that the reliability of the method was sufficient as is.

TABLE 19.1
Definitions of Fundamental Terms Applied to Validation Process

Term	Definition
Validation	Process for determination of relevance and reliability of test for a specific purpose
Reliability	Determination of reproducibility of a test within a laboratory and between laboratories (intra- and inter-laboratory variations, respectively)
Relevance	Demonstration of predictivity, applicability, and comparability of a test for a corresponding *in vivo* situation; rate-limiting factor that allows less tolerance for deviation from goal of validation
Standardization	Analysis of test procedure and its components for reliability in comparison to internal and external controls
Validity	Establishment of standards for routine intra- and inter-laboratory use of a toxicology test for purpose of attributing credibility, effectiveness, or quality
Evaluation	Process of acceptance of test by scientific community, although it does not necessarily imply achievement of this goal; appraisal of *in vitro* method without demand that process be complete
Assessment	In toxicology, appraisal of a test with respect to its value

Many of the pioneering projects that initiated the development of *in vitro* methods did not fulfill all the criteria discussed above for optimal validation (see references cited at the end of this chapter). With the exception of acute systemic toxicity studies, most laboratories used many different compounds without regard to grouping chemicals in particular classes. In some studies, however, the numbers represent mixtures of products and materials that are impossible to request for retesting, while other reports failed to present all the raw data. In many studies, *in vitro* test results were combined directly with dose-defined, gross animal toxicity data. For instance, comparisons of Draize test scores for validation against *in vitro* eye irritancy data only permitted rank correlations, thus negating a necessary *in vitro–in vivo* method of comparison (relevance). Some of the *in vitro* tests were also scored in a similar manner to rank correlations, which makes the linear correlation approach difficult.

With the exception of tests for local irritancy, validation by individual laboratories has not gained general acceptance for any other method, nor have they replaced any currently accepted analogous animal toxicology testing. As mentioned above, the explanation may reside in the lack of resources available to smaller validation programs.

Table 19.2 presents some of the early approaches to the validation process and compares them to current methods and conditions necessary for refining validation programs. Three important considerations that may determine the results of laboratory validation studies are discussed below.

19.2.2.2 Selection of Chemicals

The selection of chemicals to be used in a study is critical to the analysis of the results of any validation program. A large number of chemicals are chosen to represent as many relevant chemical and therapeutic classes. The substances are not

TABLE 19.2
Early and Current Approaches for Refining and Improving Validation Studies

Criteria	Early Approaches	Recent or Advanced Approaches
Chemicals	Few chemicals or from various classes	Large number of chemicals from relevant classes
	Water soluble and volatile chemicals; easy compatibility	Randomly selected chemicals
	Proprietary formulations	Defined chemicals from referenced databases
Data	*In vivo* data with known toxicities from various protocols	*In vivo* data with known toxicities from similar or same protocols
	Blind *in vivo* data	Coded substances
	Comparisons to animal or human toxicities	Comparisons to animal and human toxicities
Statistical analysis	Semiquantitative *in vitro* comparisons	Parallel linear comparisons
	Correlation, *ad hoc* analysis	Correlation plus toxicokinetic information
Laboratory evaluation	Single or multilaboratory evaluation of reliability	Multilaboratory evaluation of reliability and relevance
	Multicenter validation for single reliable test (multivariate analysis)	Multicenter validation for selection of battery of reliable tests (multivariate analysis)

selected based only on their physical properties in solution or their presumed cyto-toxic actions *in vivo*, for the results obtained may risk a technical bias toward the *in vitro* method. The dependable way to secure an unbiased selection is to devise a scheme for random selection from a chemical registry (e.g., Registry of Cytotoxicity, ICCCVAM, or NIOSH). A procedure validated for a limited number of classes of chemicals is probably of little practical value. Another prerequisite for a convincing validation study is to use defined chemicals [by their Chemical Abstracts Service (CAS) formulas or their generic names] rather than proprietary mixtures or coded industrial products.

Studies performed with defined chemicals allow for inter-laboratory testing and validation, thus permitting scientific legitimacy. In accordance with the selection of the chemicals and their labeling schema, presentation of the data is also important. Results of evaluations that are published without presentation of a tabulated database, for instance, risk irrelevant application and inconsistency with the target objectives.

19.2.2.3 Comparison to *In Vivo* Data

A second important factor in the validation program is the selection and comparison to analogous *in vivo* data. It is desirable to compare *in vitro* toxicity test results that have accompanying toxicokinetic parameters with *in vivo* toxicity data. Historically, earlier studies compared cell toxicity with animal LD_{50} values and found relatively good correlations, indicating comparability of cell toxicity with rodent lethality.

Few studies are concerned with the prediction of acute human toxicity by animal lethality tests or the prediction of human ocular and skin irritancy by the Draize technique. It is generally accepted that these correlations are low and that the poor comparisons are due to species differences. Ideally, in order to introduce an alternative *in vitro* test method for practical use as a procedure for screening for human toxicity, the validation step must use human toxicity data as the primary reference. The problem with human toxicity data, however, is its availability, coupled with the uneven and inconsistent quality of toxicity reference points that are available. Without a standardized database of these values, the *in vitro* methods are less reliable than the corresponding animal protocols used for comparison. Interestingly, the introduction of the EDIT program in 2000 following the successful completion of the MEIC program proposes the compilation of a human toxicity database for comparison with alternative *in vitro* methods. The program is still in progress and conclusions of *in vitro–in vivo* correlations are pending.

It is now accepted that correlations of *in vitro* and *in vivo* toxicology testing for eye irritancy for different classes of chemicals depend on toxicokinetic information available for that category. Thus, the addition of such information as the inclusion of absorption and elimination rates of chemicals in test batteries will improve future attempts at correlation. If animal LD_{50} or TD_{50} values are used as reference points, the question of the precision of the animal method, however, and the variability introduced with the species gap compounds the reliability of the cytotoxicity test. Proper use of animal data relies on comparisons to the human data for the same chemicals used in the study, thus establishing a baseline for determining whether the *in vitro* method is comparable to the animal results. Ironically, this may be construed as validation of animal tests, paradoxically included in the statistical analysis.

The bulk of available human data is currently obtained from select references and poison control centers, and is compiled from clinical case studies, hospital admissions, and emergency department visits. Although this information is not obtained systematically, it represents the most reliable source of human toxicity data available for commonly encountered chemicals. Thus, the clinical information is used as a basis for comparison against *in vitro* values. Another source of reliable human toxicology data is generated through the testing on volunteers for some types of toxicity such as skin and eye irritancy. Human skin irritation testing produces concentration effect curves for fixed endpoints, while in the case of eye irritancy, testing is limited to minimal responses (redness, itchiness).

19.2.2.4 Comparison to Other *In Vitro* Protocols

The third important factor for consideration in a validation program is a comparison to other *in vitro* data. Historically, the validation of mechanistic mutagenicity tests was based on a qualitative *in vitro–in vivo* comparison, primarily because of the difficulty associated with establishing a quantitative correlation. All *in vitro* tests of acute systemic toxicity, however, are quantitative and above all are thought to simulate quantitative interactions between chemicals and cellular components *in vivo*. The tests therefore are validated quantitatively by linear regression analysis or similar statistical methods. This analysis considers the different mechanisms of toxicity and cellular targets associated with individual chemicals and includes categorization of outlines in the design of the validation program. For example, linear regression analysis technically requires exact toxicity data without cut-off results, that is, an IC_{50} value stated to be greater than a baseline value is not meaningful for the analysis.

To date, most individual validation studies compare the results of *in vitro* toxicity tests directly with some form of animal toxicity such as acute systemic toxicity and skin or eye irritancy. Similarly, most tests are developed to account for only one of many interactions involved in the systemic toxic effects in animals including toxicity to targets not measured by the test and toxicokinetics (absorption, distribution, metabolism, and elimination). When the results from single tests are compared with general systemic toxicity, the comparison is limited to a minor part of the general toxic phenomenon measured by the test, even if this determination is not obscured by toxicokinetic phenomena. This type of validation selectively discards procedures that measure relatively rare effects and cannot discriminate between important supplementary tests and others with similar information. In addition, it is erroneous to compare *in vitro* test results to doses administered in whole animals, whether local or systemic. Instead, comparisons to tissue concentrations can be realized if toxicokinetic data are incorporated in the analysis or if the animal doses are converted to equivalent human toxicity data (see Chapter 18).

19.3 CURRENT VALIDATION PROGRAMS

19.3.1 HISTORY

The practical and economic advantages of centrally organized multilaboratory validation programs have diminished the impetus to perform individual laboratory studies. Although individual laboratory validation studies maintain a competitive edge, especially with the comparatively better flexibility afforded to smaller operations, it appears that corresponding multilaboratory approaches are more scientifically and economically feasible to satisfy regulatory requirements.

Early on, organized collaborative validation efforts enjoyed several advantages in comparison to individual laboratory studies. Such multicenter validation was believed to enhance the ideas launched by the multilaboratory evaluations of the reliability of isolated methods performed in the 1960s. In their present forms, multicenter programs consist of centrally directed testing of the same reference chemicals in many test systems and/or laboratories, with the objective of evaluating both relevance and reliability of methods. Consequently, the development of multicenter policies and procedures is analogous to other important aspects of individual laboratory validation guidelines, such as improvement in the selection of chemicals, references to *in vivo* data, and methods of correlation.

To date, many proposed alternative *in vitro* toxicology tests in the U.S. have accumulated in the scientific literature and regulatory platforms without the benefit of arriving at general acceptance. The situation is due in part to the difficult tasks and characteristics associated with introduction of systems from individual laboratories. The organization of multilaboratory validation programs is prompted by *in vitro* toxicologists as a possible remedy for the problem. In addition, public interest groups, academic institutions, industries, and government agencies in the European Union and U.S. are actively engaged in promoting the development of *in vitro* toxicology tests by organized well planned programs. Some organizations such as FRAME (Fund for Replacement of Animals in Medical Experiments, U.K.), The Johns Hopkins CAAT (Center for Alternatives to Animal Testing, Baltimore) and ERGAAT (European Research Group for Alternatives to Animal Testing) were among the original groups to coordinate efforts into proposals for future multicenter validation. The first proposed multicenter programs were outlined by FRAME. They were soon followed by special conferences on validation arranged by the CAAT group. Later, the CAAT and ERGAAT centers in collaboration with FRAME arranged a consensus conference on validation, the result of which was the development of a set of guidelines for validation of toxicity tests. The CAAT–ERGAAT conference was followed by the ERGAAT–CEC (Commission of European Communities) conference on procedures for acceptance of validated tests. These efforts prompted recognition of the importance of organized and focused multilaboratory validation programs.

19.3.2 Advantages of Organized Multilaboratory Validation Programs

Multicenter validation programs to facilitate regulatory acceptance of an alternative testing strategy have many advantages:

1. Evaluation of reliability of methods is performed in parallel with evaluation of relevance.
2. Acceptance of results improves, compared with introduction of results from individual laboratories, if an independent group not involved in assay development selects the chemicals.
3. Multilaboratory validation is considerably more economical than individual studies because efforts and resources used to select chemicals and generate *in vivo* data, including the planning and execution of comparisons, is shared among laboratories and coordinated by a central organizing body.
4. Accessible databases for validation of relevance are established, thus fostering the simultaneous evaluation of many protocols and limiting the possibility of omitting a valuable approach.
5. Organized multilaboratory studies validate the predictive ability of combined methods, facilitated by statistical multivariate analysis.
6. Organized multilaboratory programs effectively use models of various types of animal and human toxicities based on knowledge of the most critical targets for toxicity, mechanisms of action, and toxicokinetic data, and accordingly determine whether additional data are needed to contribute to these models.
7. The programs model various *in vitro* methods and data as they are received according to their applications to different types of toxicities such as acute

systemic, ocular, and skin irritancy, resulting in the discovery of distinct *in vitro* models for screening toxicity.

8. Comparison of results of different laboratories, using the same set of reference chemicals in different *in vitro* toxicity tests with different targets and criteria distinguishes differential application of contributed cytotoxicity studies.

19.3.3 ORGANIZATIONS DEDICATED TO DEVELOPMENT OF ALTERNATIVE MODELS

Several U.S., E.U, and other national and international regulatory and scientific organizations are dedicated to the development of alternative models to animal toxicology testing. Table 19.3 lists these organizations and they are further described below.

The Belgium Platform for Alternative Methods (BPAM) was established in 1997 in Brussels. Its main focus is the development and validation of alternative methods.

The British Union for the Abolition of Vivisection (BUAV) focuses on the potential of *in vitro* methods to replace *in vivo* experiments. Its report, *The Way Forward: A Non-Animal Testing Strategy for Chemicals,* contains a review of the current status of *in vitro* methods and suggests that the E.U. provide sufficient resources for the vast validation of *in vitro* methods in order to replace *in vivo* experiments.

CAAT was founded in the U.S. in 1981, with a grant from the Cosmetic, Toiletry and Fragrance Association (CTFA). Its main interests are administrative and educational promotion of techniques that replace the use of animals, reduce the need for animals for particular tests, or refine techniques in order to relieve or eliminate the pain endured by animals.

ECVAM was established in 1986. It coordinates its efforts and collaborates with several E.U. and U.S. regulatory agencies for the development and regulatory acceptance of *in vitro* tests that refine, reduce, or replace the use of laboratory animals.

ERGAAT was established in 1985 in the Netherlands. One of the earliest founding organizations for the development and evaluation of *in vitro* methods, the group is represented on the ECVAM advisory committee and is involved with ECVAM and ICCVAM related to validation of non-animal tests in regulatory toxicity testing. In addition, ERGAAT is involved in establishing risk assessment methods.

FRAME was established in the U.K. in 1969 to promote the replacement of *in vivo* experiments. In 1980, INVITTOX, a web database collection of protocols for *in vitro* methods, was established by FRAME and is now part of ECVAM's Scientific Information Service. FRAME participated in the validation study of the 3T3 NRU phototoxicity test in 1997 that was accepted by the E.U. in 2000, and is undergoing validation by ICCVAM–NICEATM (see below). It also sponsors and publishes a leading toxicology journal for the development of alternative models, *Alternatives to Laboratory Animals (ATLA).*

The Interagency Coordinating Committee on the Validation of Alternative Methods (ICCVAM) was established in 1997 in the U.S. as a standing committee by the director of the National Institute of Environmental Health Sciences (NIEHS). By 2000, it received permanent status through Congressional authorization. As a regulatory agency, its function covers the organization, development, and validation of

TABLE 19.3
Regulatory and Scientific Organizations Dedicated to Development of Alternative Models to Animal Toxicology Testing

Organization*		Representative Country	Description
BPAM	Belgium Platform for Alternative Methods	Belgium	Regulatory, administrative
BUAV	British Union for the Abolition of Vivisection	U.K.	Regulatory, administrative
CAAT	John Hopkins Center for Alternatives to Animal Testing	U.S.	Academic
ECVAM	European Center for Validation of Alternative Methods	E.U.	Regulatory
EDIT	Evaluation-Guided Development of *In Vitro* Tests	Sweden	Multilaboratory program
ERGAAT	European Research Group for Alternatives in Toxicity Testing	E.U.	Regulatory
FRAME	Fund for the Replacement of Animals in Medical Experiments	U.K.	Private foundation
ICCVAM	Interagency Coordinating Committee on Validation of Alternative Methods	U.S.	Regulatory; multilaboratory program
IIVS	Institute for *In Vitro* Sciences	U.S.	Private, non-profit
IVSS**	*In Vitro* Specialty Section of U.S. Society of Toxicology	U.S.	National toxicology society
JaCVAM	Japanese Center for Validation of Alternative Methods	Japan	Regulatory; multilaboratory program
MEIC	Multicenter Evaluation of *In Vitro* Cytotoxicity	Sweden, Finland, Norway, Denmark	Multilaboratory program
NCA	Netherlands Center for Alternatives to Animal Use	Netherlands	Regulatory, administrative
NICA	Nordic Information Center for Alternative Methods	Sweden	Coordination of EDIT project
ZEBET	Center for Documentation and Evaluation of Alternative Methods to Animal Experiments	E.U.	Regulatory
ZET	Center for Alternative and Complementary Methods in Animal Experiments	Austria	Administrative

* Websites cited in reference list.
** Name has been officially changed to IVAM — In Vitro and Alternative Methods Specialty Section.

alternative methods as well as the establishment of criteria and processes for the regulatory acceptance of *in vitro* methods. In 1998, the National Toxicology Program (NTP) Interagency Center for the Evaluation of Alternative Toxicological Methods (NICEATM) was established. ICCVAM and NICEATM work collaboratively along with scientific and public advisory committees such as the Scientific Advisory Committee for Alternative Toxicological Methods (SACATM-ICCVAM), for the establishment of methods that are predictive for adverse human toxicological effects.

The Institute for *In Vitro* Sciences (IIVS) was founded in 1997 in the U.S. The goals of the non-profit organization center on evaluation of testing for skin and ocular toxicity, bioassay development, and the validation of *in vitro* methods.

The *In Vitro* Specialty Section (IVSS) of the U.S. Society of Toxicology was founded in 1994. This special internal group is interested in the application of *in vitro* techniques to problems of cellular toxicity, understanding basic cellular processes involved in the induction of adverse outcomes of specific organs and the whole animal, and the development of cellular and subcellular systems to predict toxicity *in vivo* for risk assessment purposes. The group also supports *in vitro* test validation and all aspects of test development and acceptance for scientific or regulatory purposes.

The Multicenter Evaluation of *In Vitro* Cytotoxicity (MEIC) was initiated by Björn Ekwall in 1989 and organized by a committee of the Scandinavian Society for Cell Toxicology (SSCT). From 1989 to 1996, international laboratories voluntarily participated in testing 50 representative chemicals with known and varied human and animal toxicities. The continuing and final results of MEIC studies were published in a series of reports indicating that IC_{50} values from 68 independent cytotoxicity assays in combination with toxicokinetic data predicted human lethal dosages as efficiently as rodent LD_{50} values. These studies also suggested that primary cultures and human cell lines are more sensitive to chemical toxicity than continuous animal cell lines.

By 1998, the MEIC project was extended into the Evaluation-Guided Development of *In Vitro* Tests (EDIT) in order to organize global collaboration of laboratories for the establishment of new *in vitro* tests. The EDIT project incorporates toxicokinetics studies in order to obtain a battery of tests for tier testing in animal toxicology. The Nordic Information Center for Alternative Methods (NICA) coordinates the EDIT project and was established in Sweden to promote the use of *in vitro* tests for regulatory toxicology.

The Netherlands Center for Alternatives to Animal Use (NCA) was established in 1994. It focuses on the development, acceptance, and promotion of alternative methods in the Netherlands.

The Center for Documentation and Evaluation of Alternative Methods to Animal Experiments (ZEBET) was established in Germany in 1989 to reduce and replace animal experimentation with alternative models. ZEBET provides a database of 347 chemicals for *in vitro* cytotoxicity data that includes their corresponding *in vivo* acute toxicity data. Known as the Registry of Cytotoxicity (RC) database, the information is used to develop regression models to determine starting doses for *in vivo* lethality. Moreover, ZEBET undertakes validation studies of *in vitro* methods in cooperation with ECVAM and the European Cosmetic, Toiletry and Perfumery Industry (COLIPA).

The Center for Alternative and Complementary Methods in Animal Experiments (ZET) was established in 1996 in Austria with the aim to develop, improve, and validate alternative methods using the "3R" (replacement, refinement, and reduction) approach. ZET has released a series of publications based on its sponsored international congresses and discussing alternative methods to animal testing.

19.3.4 ORGANIZED PROGRAMS

Current or completed multilaboratory validation studies of alternative *in vitro* ocular and dermal toxicology testing are discussed in Chapter 13. These and other second generation validation studies are facilitated through collaborative international cooperation, particularly through the efforts of ICCVAM, NICEATM, ECVAM, and ZEBET. These collaborations provide a foundation for validating alternative methods along with promotion and encouragement for the harmonization of scientific approaches to validation and review. In addition, both ICCVAM and ECVAM have outlined key objectives: to stimulate development of test methods and testing strategies in prioritized areas by individual laboratories and to facilitate the nomination of promising test methods for appropriate regulatory translation. Table 19.4 summarizes current *in vitro* validation programs, proposals, test method evaluations, and testing strategies currently sponsored by ICCVAM.

19.4   SUMMARY

Current and future validation should not be hampered by theoretical considerations of the best way to achieve regulatory acceptance. The refinement of the validation programs should be approached as a set of generally applicable rules. The most effective future validation will probably require studies directed at all levels, including the efforts of individual laboratories, reliability-oriented multilaboratory programs, and multiple-method multilaboratory perspectives focused on cell based toxicology.

Combinations of different objectives will be evaluated in different studies. For instance, several methods could be analyzed for reliability in some programs, while simultaneously evaluated for specific or general relevance in other programs. It may not be possible to combine some objectives such as selection of chemicals, choice of *in vivo* data, and demands on correlative methods in a single study. Human data on acute systemic toxicity may not be compatible with particular selections of certain chemicals. Some conflicts of goals may arise in the selection of optimum objectives in the scheme of validation proposals.

It is important to understand that the progress of *in vitro* cytotoxicology testing relies on the continuous development of better strategies to evaluate new tests rather than as a set of inflexible guidelines for the inclusion or exclusion of protocols. Thus, efforts to evaluate the scientific integrity of methods for validation purposes should be administered by independent investigative groups rather than government sponsored organizations. The groups must be free of vested interests, with their services directed toward the development of scientifically valid programs.

TABLE 19.4
Current or Completed Validation Proposals, Programs, and Evaluation of Testing Strategies and Methods Considered by ICCVAM-NICEATM*

Description	NIH Publication No.	Year
Validation study of *in vitro* test for acute systemic toxicity; joint effort of NICEATM and ECVAM	Currently undergoing laboratory evaluation	
Independent scientific peer review of use of *in vitro* testing methods for estimating starting doses for acute oral systemic toxicity tests	—	2006
Current status of *in nitro* test methods for detecting ocular corrosives and severe irritants	—	2006
Recommended performance standards for *in vitro* test methods for skin corrosion	04-4510	2004
Evaluation of *in vitro* test methods for detecting potential endocrine disruptors: estrogen receptor and androgen receptor binding and transcriptional activation assays	03-4503	2003
ICCVAM evaluation of Episkin™, EpiDerm™, and rat skin transcutaneous electrical resistance (TER) assay; *in vitro* test methods for assessing dermal corrosivity potentials	02-4502	2002
Revised up-and-down procedure: test method for determining acute oral toxicity	02-4501	2001
FETAX [frog (Xenopus) embryo teratogenesis assay]	—	2000
Murine local lymph node assay: test for assessing allergic contact dermatitis potentials	99-4494	1999
Corrositex™ *in vitro* test for assessing dermal corrosivity potentials	99-4495	1999

* See http://iccvam.niehs.nih.gov/docs.

SUGGESTED READINGS

Arias-Mendoza, F. et al., Methodological standardization for a multi-institutional *in vivo* trial of localized 31P MR spectroscopy in human cancer research: *in vitro* and normal volunteer studies, *NMR Biomed.*, 17, 382, 2004.

Balls, M. and Combes, R., The need for a formal invalidation process for animal and non-animal tests, *Altern. Lab. Anim.*, 33, 299, 2005.

Combes, R., Barratt, M., and Balls, M., An overall strategy for the testing of chemicals for human hazard and risk assessment under the EU REACH system, *Altern. Lab. Anim.*, 34, 15, 2006.

Gartlon, J. et al., Evaluation of a proposed *in vitro* test strategy using neuronal and non-neuronal cell systems for detecting neurotoxicity, *Toxicol. in Vitro*, 20, 1569, 2006.

Hoffmann, S. and Hartung, T., Designing validation studies more efficiently according to the modular approach: retrospective analysis of the Episkin test for skin corrosion, *Altern. Lab. Anim.*, 34, 177, 2006.

Prieto, P. et al., The assessment of repeated dose toxicity in vitro: a proposed approach: report and recommendations of ECVAM workshop 56, *Altern. Lab. Anim.*, 34, 315, 2006.

Spielmann, H. et al., The practical application of three validated *in vitro* embryotoxicity tests: report and recommendations of an ECVAM/ZEBET workshop, *Altern. Lab. Anim.*, 34, 527, 2006.

REVIEW ARTICLES

Barile, F.A., Continuous cell lines as a model for drug toxicity assessment, in *Introduction to In Vitro Methods in Pharmaceutical Research*, Castell, J.V. and Gómez-Lechón, M.J., Eds., Academic Press, London, 1997.

Barile, F.A., Dierickx, P.J., and Kristen, U., *In vitro* cytotoxicity testing for prediction of acute human toxicity, *Cell Biol. Toxicol.*, 10, 155, 1994.

Clemedson C. et al., MEIC evaluation of acute systemic toxicity I. Methodology of 68 *in vitro* toxicity assays used to test the first 30 reference chemicals, *Altern. Lab. Anim.*, 24, 249, 1996.

Clemedson, C. et al., MEIC evaluation of acute systemic toxicity II. *In vitro* results from 68 toxicity assays used to test the first 30 reference chemicals and a comparative cytotoxicity analysis, *Altern. Lab. Anim.*, 24, 273, 1996.

Clemedson, C. et al., MEIC evaluation of acute systemic toxicity III. *In vitro* results from 16 additional methods used to test the first 30 reference chemicals and a comparative cytotoxicity analysis, *Altern. Lab. Anim.*, 26, 93, 1998.

Clemedson, C. et al., MEIC evaluation of acute systemic toxicity IV. *In vitro* results from 67 toxicity assays used to test reference chemicals 31–50 and a comparative cytotoxicity analysis, *Altern. Lab. Anim.*, 26, 131, 1998.

Clemedson, C. et al., MEIC evaluation of acute systemic toxicity VII. Prediction of human toxicity by results from testing of the first 30 reference chemicals with 27 further *in vitro* assays, *Altern. Lab. Anim.*, 28, 161, 2000.

Curren, R., Bruner, L., Goldberg, A., and Walum E., Validation and acute toxicity testing: 13th meeting of Scientific Group on Methodologies for the Safety Evaluation of Chemicals, *Environ. Health Perspect.*, 106, 419, 1998.

Edler, L. and Ittrich, C., Biostatistical methods for the validation of alternative methods for *in vitro* toxicity testing, *Altern. Lab. Anim.*, 31, 5, 2003.

Ekwall, B., Overview of final MEIC results II. The *in vitro–in vivo* evaluation including the selection of a practical battery of cell tests for prediction of acute lethal blood concentrations in humans, *Toxicol. in Vitro*, 13, 665, 1999.

Ekwall, B. and Barile, F.A., Standardization and validation, in *In Vitro Cytotoxicology: Mechanisms and Methods*, CRC Press, Boca Raton, FL, 1994, p. 189.

Ekwall, B. et al., MEIC evaluation of acute systemic toxicity V. Rodent and human toxicity data for the 50 reference chemicals, *Altern. Lab. Anim.*, 26, 569, 1998.

Halle, W., Spielmann, H., and Liebsch, M., Prediction of human lethal concentrations by cytotoxicity data from 50 MEIC chemicals, *ALTEX*, 17, 75, 2000.

Hellberg, S. et al., Analogy models for prediction of human toxicity, *Altern. Lab. Anim.*, 18, 103, 1990.

Interagency Coordinating Committee on the Validation of Alternative Methods, Biennial Progress Report of the National Toxicology Program, Publication 04-4509, National Institute of Environmental Health Sciences, Research Triangle Park, NC, 2003. http://iccvam.niehs.nih.gov/about/annrpt/bienrpt044509.pdf.

Louekari, K., Status and prospects of *in vitro* tests in risk assessment, *Altern. Lab. Anim.*, 32, 431, 2004.

Paris, M. et al., Phase I and II results of a validation study to evaluate *in vitro* cytotoxicity assays for estimating rodent and human acute systemic toxicity, 43rd Annual Meeting of Society of Toxicology, *Toxicol. Sci.*, Abstract 240, 2004.

Rispin, A., Stitzel, K., Harbell, J., and Klausner, M., Ensuring quality of *in vitro* alternative test methods: current practice, *Regul. Toxicol. Pharmacol.*, 45, 97, 2006.

Spielmann, H., Genschow, E., Leibsch, M., and Halle W., Determination of starting dose for acute oral toxicity (LD_{50}) testing, *Altern. Lab. Anim.*, 27, 957, 1999.

Ubels, J.L. and Clousing, D.P., *In vitro* alternatives to the use of animals in ocular toxicology testing, *Ocul Surf.*, 3, 126, 2005.

20 Applications of Alternative Models for Toxicology Testing

20.1 INTRODUCTION

In the past, the only systematic attempts to develop alternative models for animal toxicology testing were commandeered by academic endeavors aimed at conducting basic research, supplemented by their efforts at soliciting industrial support for launching *in vitro* assays. In the main, regulatory agencies displayed a cavalier attitude toward the field, thus discounting the importance of changing perceptions within the pharmaceutical, cosmetic, and toxicology testing industries.

As increasing pressure has mounted from the scientific and public interest groups, the field of alternative *in vitro* methods has witnessed favorable transformations and developments including applications of the technology and its introduction into the regulatory arena. Although it is arguable that progress is unhurried and not extensive, consider the historical status of *in vitro* methods 20 years ago and the formidable obstacles that have been overcome. Several developments in the regulatory arena are presented as evidence of significant advances in this direction including (1) the promulgation of OECD regulations in the European Union, (2) the establishment of ICCVAM, NICEATM, and ECVAM agencies and the SACATM advisory committee, (3) the founding of international societies for fostering the development of alternative methods, and (4) advances arising from workshops and validation programs.

20.1.1 PROMULGATION OF OECD AND FRAME REGULATIONS BY THE EUROPEAN UNION

The forerunner of the OECD was the Organization for European Economic Co-operation (OEEC) that was formed to administer American and Canadian aid under the Marshall Plan for reconstruction of Europe after World War II. Since 1961, OECD's goals have been the building of strong economies in its member countries and promoting regulatory and economic reforms in industrialized and developing nations. It also fosters policy recommendations, country reviews, thematic discussions, and cooperation with non-member countries.

OECD introduces and promotes guideline lists and reports based on physical chemistry, ecotoxicology, degradation and accumulation, and short- and long-term toxicology testing by expert groups. It also publishes a series of documents on good laboratory practices* to ensure that *in vitro* tests used for regulatory purposes are reproducible, credible, and acceptable.

Other efforts guided by the experiences from validation studies for alternative methods in toxicology and OECD initiatives include concepts of good cell culture practice (GCCP) that aim to define minimum quality standards for *in vitro* techniques.

The Fund for the Replacement of Animals in Medical Experiments (FRAME) is a registered U.K. charity whose objective is the promotion of alternatives to animal use in research, testing, and education. While it considers that the current scale of animal experimentation is unacceptable, it recognizes the need for adequately testing new consumer products, medicines, and industrial and agricultural chemicals to identify potential hazards to human and animal health. Consequently it advocates the "3R" approach (replacement, refinement†, and reduction) to animal experimentation and seeks to promote a moderate approach by encouraging realistic considerations of the ethical and scientific issues involved in animal toxicology testing.

FRAME recently introduced a number of suggestions for improving the initiatives of the European Union REACH (registration, evaluation, and authorization of chemicals) system for safety testing and risk assessment, first proposed as a White Paper in 2001. These initiatives considered the scientific and animal welfare issues raised by the REACH proposals and resulted in adoption of legislative actions through the European Council and European Parliament.

20.1.2 ESTABLISHMENT OF ICCVAM, NICEATM, AND ECVAM REGULATORY AGENCIES

Over the past decade, based on provocation and encouragement from scientific and public interest groups, regulatory authorities in the United States and European Union launched initiatives to validate new and improved toxicological test methods. In the U.S., the Interagency Coordinating Committee on the Validation of Alternative Methods (ICCVAM) and its supporting National Toxicology Program (NTP) Interagency Center for the Evaluation of Alternative Toxicological Methods (NICEATM) were established by the federal government to work with test developers and federal agencies to facilitate the validation, review, and adoption of new scientifically sound test methods including alternatives that refine, reduce, and replace animal use.

ICCVAM was established in 1997 according to the ICCVAM Authorization Act of 2000‡ by the director of the National Institute of Environmental Health Sciences (NIEHS) to implement NIEHS directives.** It is a permanent committee composed

---

* Advisory document of Working Group on Good Laboratory Practice, The Application of the Principles of GLP to *In Vitro* Studies, OECD Series on Principles of Good Laboratory Practice and Compliance Monitoring, 2004.
† Revised definition includes lessening or eliminating pain.
‡ Public Law (P.L.) 106-545.
** P.L. 103-43 directed NIEHS to develop and validate new test methods and establish criteria and processes for the validation and regulatory acceptance of toxicological testing methods.

of representatives from 15 federal regulatory and research agencies and advisory groups. The committee coordinates cross-agency issues relating to development, validation, acceptance, and national and international harmonization of toxicology test methods.

NICEATM was established in 1998 to provide operational support for ICCVAM and carry out committee-related activities such as peer reviews and workshops for test methods of interest to federal agencies. NICEATM and ICCVAM coordinate scientific reviews of the validation statuses of proposed methods and provide recommendations regarding their usefulness to appropriate agencies. Together, the agencies promote the validation and regulatory acceptance of toxicology test methods that are more predictive of adverse human and ecological effects than currently available methods and refine, reduce, and replace animal use.

ECVAM was created in the European Union in 1992 by a communication from the Commission to the Council and the Parliament in October 1991* that requires the commission and member states to actively support the development, validation, and acceptance of methods that can reduce, refine, or replace the use of laboratory animals. ECVAM's duties include:

1. Coordinating the validation of alternative test methods at European Union level
2. Promoting exchanges of information on the development of alternative test methods
3. Generating, maintaining, and managing a database of alternative procedures
4. Encouraging dialogs among legislators, industries, biomedical scientists, consumer organizations and animal welfare groups, with a view to the development, validation, and international recognition of alternative test methods

Both ICCVAM and ECVAM have adopted similar validation and regulatory acceptance criteria for the adoption and use of alternative test methods. They have fostered and encouraged international collaborations to facilitate the adoption of new test methods by involving sharing of expertise and data for test method workshops and independent scientific peer reviews. Among their recent activities, NICEATM and ECVAM initiated a joint international validation study on *in vitro* methods for assessing acute systemic toxicity. These collaborations are expected to contribute to accelerated international adoption of harmonized new test methods that support improved public health and provide for reduced and more humane use of laboratory animals.

The feasibility and potential benefits of such a partnership between two large regulatory organizations continue to evolve as the development of alternative methodologies progresses. The fear, however, is that this extraordinary increase in the introduction of test methods from the scientific and industrial communities and the accompanying pressure to deliberate on and implement the procedures may strain the capacities of such institutions. Consequently, coordination of functions, maintenance of scientific standards, and commitments to the principles of validation support

---

* The communication is a requirement in Directive 86/609/EEC on the protection of animals used for experimental and other scientific purposes.

TABLE 20.1
Additional Organizations Sponsoring Activities in Support of Humane Treatment of Animals

	Organization	Origin
ARDF	tives Research and Development Foundation	Jenkintown, PA
EUROTOX	Federation of European Toxicologists and European Societies of Toxicology	E.U.
FRAME	Fund for Replacement of Animals in Medical Experiments	U.K.
HSUS	Humane Society of the United States	Washington, D.C.
IFER	International Foundation for Ethical Research	Chicago, IL
IVSS*	*In Vitro* Specialty Section of the U.S. Society of Toxicology	Reston, VA
SSCT	Scandinavian Society for Cell Toxicology	Sweden
WCAAUFS	World Congress on Alternatives and Animal Use in the Life Sciences	E.U

* Name has been officially changed to In Vitro and Alternative Methods (IVAM) Specialty Section.

opportunities for efficient execution of animal welfare objectives and the implementation of alternative tests on the regulatory agenda.

20.1.3 ESTABLISHMENT OF INTERNATIONAL EFFORTS FOR DEVELOPMENT OF ALTERNATIVE METHODS

Several other supportive agencies, societies, and congresses have assumed more active roles, particularly with financial, political, and scientific encouragement to support the humane treatment of animals. Some of these organizations are listed in Table 20.1.

Among these, the *In Vitro* Specialty Section (IVSS) of the Society of Toxicology (SOT) in the U.S. is composed of members with expertise in the application of *in vitro* techniques to problems of cellular toxicity, with a special emphasis on safety evaluation. The interests of the IVSS include studies of the basic cellular processes involved in the induction of adverse outcomes of specific organs and the whole animal and the development of simple to complex cellular and subcellular systems to predict toxicity *in vivo* and to assess risk. The IVSS holds regular meetings at the annual SOT congress to discuss topics of interest.

20.1.4 WORKSHOPS AND VALIDATION PROGRAMS

The ECVAM validation concept defined at two validation workshops held in Amden, Switzerland, in 1990 and 1994 takes into account the essential elements of pre-validation and prediction models based on biostatistics criteria. The program has been accepted officially by E.U. member states, OECD, and the regulatory agencies of the U.S. The ECVAM validation concept was introduced into the ongoing ECVAM/COLIPA validation study (see Chapter 19) of *in vitro* phototoxicity tests that ended successfully in 1998.

The 3T3 neutral red uptake *in vitro* phototoxicity test was the first experimentally validated *in vitro* toxicity test recommended for regulatory purposes by the ECVAM Scientific Advisory Committee (ESAC) and was accepted into E.U. legislation for testing of new chemicals in 2000. From 1996 to 1998, two *in vitro* skin corrosivity tests were successfully validated by ECVAM and were also officially accepted into E.U. regulations in 2000. From 1997 to 2000, an ECVAM validation study on three *in vitro* embryotoxicity tests was successfully completed. The three tests [whole embryo culture (WEC) test, the micromass test on limb bud cells, and the embryonic stem cell test (EST)] are encouraged for routine use in laboratories of the European pharmaceutical and chemical industries. Meanwhile, in 2002, the OECD Test Guidelines Program was considering the worldwide acceptance of the validated *in vitro* phototoxicity and corrosivity tests.

In the U.S., some research laboratories, while trying to comply with the regulations established by the U.S. Animal Welfare Act and its amendments (1985), have been forced to terminate their efforts because of the high cost of maintaining and providing for animals as required by the act. In response, the U.S. Public Health Service established the Office of Animal Research and the Office of Science Education to promote science education, coordinate policies, and distribute information on the use of animals in research. Historically, the use of animals in biomedical research has provided and will continue to provide valuable information about chemicals, drugs, and products at the vanguard of public health. Consequently, the burden of developing suitable protocols to test the safety and efficacy of thousands of drugs and chemicals used in medicine or introduced by the chemical industry is incumbent upon the scientific community.

20.2 SIGNIFICANCE OF ALTERNATIVE MODELS FOR ANIMAL TOXICOLOGY TESTING

20.2.1 In Vitro Tests and Significance

No single method is expected to cover the complexity of general toxicity in humans or animals. The response of the mammalian organism to a chemical is known to involve various physiologic targets and a variety of complex toxicokinetic factors. Chemicals induce injury through an assortment of toxic mechanisms. In addition, a number of questions have emerged as a result of previous and current validation studies of the prediction systemic and local toxicity by *in vitro* tests:

1. Are alternative methods capable of fulfilling the need to model different types of quantitative general toxicity such as acute systemic toxicity or local irritancy? This goal may require several relevant systems such as human hepatocytes, heart, kidney, lung, nerve cells, and other cell lines of applicable and relevant importance.
2. Are the results generated from these methods predictive of basal cytotoxicity* in general and local toxicity in particular?

---

* Toxicity of a chemical that affects basic cellular functions and structures common to all human and animal specialized cells regardless of organ of origin.

3. Are quick, simple, and economic *in vitro* cell systems capable of sustaining the pressure and standards of safety and regulation demanded by the needs of public health?

20.2.2 FUNCTIONAL CLASSIFICATION OF CYTOTOXICITY BASED ON *IN VITRO* METHODOLOGIES

Other than the ability of a chemical to possess genotoxic or carcinogenic potential, cellular toxicity is classified in several ways. One example is the distinction between local and systemic toxicity. Other classifications include, but are not limited to (1) acute versus chronic toxicity (see below), (2) immediate versus delayed toxicity, (3) diffuse versus targeted toxicity, and (4) high dosage versus low dosage toxicity. Figure 20.1 depicts the functional and cytotoxic classifications of primary toxic mechanisms based on different *levels of physiologic organization* affected by chemicals. The ability of a toxic substance to target functional areas correlates this level with the corresponding cytotoxic classification.

20.2.2.1 Basal Cell Functions

These fundamental processes involve structures and functions common to all cells in the human body and include the processes necessary to maintain cell membrane integrity, mitochondrial oxidation, translation and transcription, and lysosomal enzyme activity. In cultured cells, these functions are generally delegated to undifferentiated or dedifferentiated cell lines that are maintained in continuous culture over several passages. Some cell lines do not require several passages but may dedifferentiate after several hours in culture. Hepatocytes, for instance, lose their high initial levels of enzyme activities after several hours in culture. These cell lines, however, are not necessarily without specific functional characteristics. Thus, most cell lines of mesenchymal origin (fibroblasts), for instance, maintain their protein secretory activities, although many of their original morphological and biochemical features are lost.

It must be remembered that these types of continuous cells have been traditionally used for mechanistic toxicological and pharmacological studies. Thus, regardless

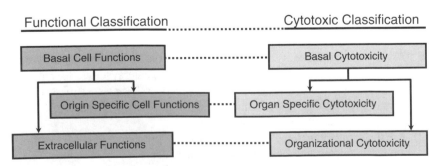

FIGURE 20.1 Classification of chemicals based on functional and cytotoxic levels of physiologic organization.

of their organs of origin, finite or immortal dedifferentiated continuous cell lines are grouped into different toxicological and pharmacological categories because of their varied *mechanistic* responses to chemicals. Cell lines with specialized persistent metabolic functions such as MDCK kidney cells or neuroblastoma 1300 cells demonstrate reactions to toxic agents based on specialized functions that persist to some extent. In cytotoxicity testing, however, most chemicals precipitate the same toxic alterations of basal cell functions in cell types of different origin as well as in cells that are similar to those lacking most cell-specific functions.

20.2.2.2 Origin-Specific Cell Functions

These structures and functions distinguish cells of varied origin and are classified as specialized cells with specific cell functions. These include structures that are unique to the type of cell, help identify the cell type, and may require an energy source in addition to their basal functions. Examples of such structures and functions include the presence of cilia (tracheal epithelial cells), contractility (myocardial cells), hormone production (cells of glandular or intestinal origin), and cells with high enzymatic activity capable of demonstrating chemical detoxification (hepatocytes or lung cells of epithelial origin).

20.2.2.3 Extracellular Functions

Certain cell pathways produce soluble mediators destined for extracellular activities. Their role in cytotoxic classification, therefore, is limited primarily because interference with their activity at the destination does not reflect their cell origins. Secretory processes such as the release of neurotransmitters by neurons and secretion of collagen by lung fibroblasts or cytokines by cells of immunity are difficult to trace to a target organ or cell within a cell culture system. The synthesis of extracellular matrix components is another example in which interference by a toxic agent at the sight of deposition does not reflect the effect of that chemical on epithelial cells. Of greater value as an indicator of toxicity is interference with these extracellular components by chemical substances either by non-specific binding to the component or blocking receptors on the target cell. Nevertheless, the measurement of these functions *in vitro* overlaps with origin-specific functions, and thus may be an indication of organ-specific cytotoxicity (see below).

20.2.3 CYTOTOXIC CLASSIFICATION BASED ON *IN VITRO* METHODOLOGIES

Analogous to functional classification, there are three types of mechanisms for toxicity and thus a parallel system for cytotoxic classification is described.

20.2.3.1 Basal Cytotoxicity

Basal cytotoxicity is an assault upon fundamental processes that involves structures and functions common to all cells in the body including cell membranes, mitochondria, ribosomes, chromosomes, and lysosomes. For instance, a chemical that inhibits cell proliferation at low or moderate concentrations is considered to have basal cytotoxic effects. Alternatively, a mutagen directly alters DNA sequences or interferes with chromosomal arrangement and thus also has the potential for basal cytotoxicity.

This concept should not be confused with the traditional ideas of systemic toxicity that excludes the contributions of genotoxicity to general systemic activity. In actuality, it is conceivable that a mutagen may act in a basal cytotoxic fashion because of chemical interference with a fundamental structure common to all cells such as a chromosome.

Basal cell functions generally support organ-specific cell functions. Chemicals capable of affecting basal cell activities are also likely to interfere with specialized functions. Consequently, studies aimed at understanding basal cytotoxic phenomena are designed with either primary cultures or continuous cell lines. Continuous cultures offer the advantages of working with more homogeneous populations, are well characterized, and are more easily manipulated in comparison to primary cultures. Among continuous maintenance cultures, embryonic or adult immortal cell lines are easily maneuvered and afford longer *in vitro* life spans. Although many continuous cell lines retain highly specialized biological activities *in vitro* that are characteristic of their original tissues, significant loss of differentiation occurs with continued passages.

Basal cytotoxicity from a chemical, therefore, is demonstrated as a result of chemical insult upon cell membrane integrity, mitochondrial activity, or protein or DNA synthesis, since these are examples of fundamental metabolic functions common to all cells. In fact, the metabolic activities of cultured cells reflect their adaptation to *in vitro* conditions rather than their primary sites of origin, which may account for different cell types displaying similar responses to toxins. The level of activity in these cells, however, varies from one cell type to another, depending on the cell's metabolic rate and contribution to homeostatic mechanisms, but the processes do not differ qualitatively among organs. If chemical injury in an animal or human is a representation of the failure of the entire system to support the target organs, then all cells will display similar dysfunctions in their ability to maintain homeostasis. In many cases, target organ toxicity in humans is caused by basal cytotoxicity of chemicals distributed to the corresponding organ. If this hypothesis is true, tests using primary cultures and continuous differentiated cell lines could be developed to cover a large percentage of toxic effects and reduce the need to introduce many laborious systems with organ-specific cells that are structurally and mechanistically unrelated.

In order to show that a response to a toxic agent in a cell line is an expression of a basal cytotoxic phenomenon, several criteria must be met:

1. A substantial number of chemicals of various categories should be tested and shown to induce a degree of toxicity at moderate concentrations, especially those that are known to be toxic *in vivo*.
2. Indeed, metabolic functions of various cell lines removed from parent cells show similar responses to toxic agents, evidence of which is seen in several multi-center studies (see reference list at end of this chapter).
3. Cytotoxic agents with similar mechanisms that show the same cytotoxic response *in vivo* and *in vitro* support the contention that this reaction is due to basal cytotoxicity.

Such experimental designs have also recently been incorporated into multicenter studies and form the basis of current validation programs. The use of cultured cells for toxicity testing and screening has not progressed without objections. Among these, the most important are based on current technical limitations that have not been fully investigated.

Some types of cells require hormonal, neuronal, and immunological adjustments, that are not readily present in routine monolayer cultures. Models designed to mimic extracellular functions and organizational toxicity, as with the incorporation of co-cultures and culture inserts, may improve the ability to distinguish among the three levels of cytotoxicity.

Many continuous cell lines are not capable of metabolizing xenobiotics. As a result, the effects of chemicals that require biotransformation for their toxicity will not be observed. The use of organ-specific primary cultures that retain high levels of enzymatic activity may overcome this drawback.

The significance of exposure time has not been adequately addressed. At present, validated tests and tests undergoing evaluation measure cytotoxicity according to three types of exposure: (1) short-term tests involving exposure for periods up to 72 hr, (2) long-term tests, for periods of 4 to 7 days, and (3) special tests involving unstable, volatile, or insoluble materials. The criticisms imply that the exposure periods represent short incubation and observation times and are not predictive for acute *in vivo* effects. On the contrary, cell culture systems are modeled to mimic chronic exposure *in vivo* through manipulation of subcultivation parameters (see Section 20.2.4).

20.2.3.2 Organ-Specific Cytotoxicity

Chemicals that exhibit selective toxicity, especially when *in vitro* concentrations are lower in comparison to *in vivo* concentrations, qualify for organ-specific toxicity — the effect of a chemical on a specialized function characteristic of a particular cell type, that is, origin-specific cell function. Measurement of these cell functions and extrapolation to organ-specific toxicity requires the use of primary cultures that have morphological and biochemical features similar to those of the original tissue. Therefore they offer the possibility for comparative studies of specialized tissues obtained from a variety of mammalian species.

Primary cultures are generally more sensitive to the effects of toxic chemicals than cell lines because of their simultaneous role of adapting to the cell culture environment while influenced in the presence of a toxic substance. The main limitations of primary cultures include the difficulty of technically establishing the culture, the low cell yields, the lack of homogeneity, and the rapid loss of organ and cell-specific markers that determine organ-specific cytotoxicity.

Depending on the cell type, the use of primary cultures assures maintenance of specialized functional markers present in the parent cell. Thus, by comparing similar classes of chemicals against primary and continuous cultures from the same organ, the differentiation of basal from organ-specific cytotoxicity is realized. In addition, primary and continuous cultures from the same organ incubated with structurally similar chemicals are used to distinguish a substance with a predilection for a particular organ. For example, in order to screen hepatotoxic agents, it is necessary to use a combination of primary hepatocytes and parallel cultures of dedifferentiated cells. If a considerably lower concentration of a substance is needed to inhibit 50% of proteins synthesized (IC_{50}) in the primary hepatocyte culture than in an analogous parenchymal hepatoma cell line, this suggests that the chemical is mechanistically hepatotoxic. Similar moderate concentrations of a particular chemical that cause a toxic response in both types of cultures could be explained as a basal cytotoxic phenomenon.

20.2.3.3 Organizational Cytotoxicity

Chemicals that obstruct metabolic or secretory pathways interfere with both organ-specific functions and organizational functions. Some examples of chemicals or drugs that cause organizational toxicity include receptor antagonists (muscarinic blockers such as atropine), immune modulators (immunosuppressants or inhibitors of cytokine release), or inhibitors of neurotransmitter metabolism (selective serotonin reuptake inhibitors). Therefore, *in vitro* tests to screen for these types of organizational toxicity incorporate systems capable of detecting the passage or movement of molecules throughout the media or across membranes. Models such as co-cultures in filter inserts are used to predict the toxicities of chemical metabolites or secretory products of cells and their effects on the apposed monolayer. Models for intestinal absorption employ this system. In addition, agents capable of preventing the binding of macromolecules with specific cell receptors qualify as interfering with organizational toxicity.

These classifications are important for understanding the possibilities of cellular methods in toxicology testing. Basal cytotoxic mechanisms are studied in systems with undifferentiated finite or continuous cell lines, while organ specific cytotoxicity is analyzed in primary cultures with well differentiated cells from different organs. Organizational toxicity is observed indirectly in cell cultures by examining the substrates or products of cell metabolism. It is important to note, however, that basal functions in cells are prerequisites for specific functions basal cytotoxicity *imitates* the effects of organ specific cellular toxicity or even organizational toxicity by indirectly affecting metabolic functions. Also, the mode of toxicity of a chemical is specific *within* each type of toxicity. For example, basal cytotoxicity precipitated by a lipophilic chemical can act either by a specific action on the sodium ion pump or as a result of a non-specific dissolution of the chemical to the non-polar regions of the cell membrane. The net result of both mechanisms is loss of membrane integrity.

20.2.4 Measurements of Acute and Chronic Toxicity

A major criticism of *in vitro* cytotoxicity assays is that the procedures measure only acute toxicity, primarily because the exposure is of short duration and occurs through one cellular passage level. This is because occasionally it is difficult to determine the difference between acute and chronic exposure in culture. Since one culture cycle (one passage level) can extend from 3 to 7 days, depending on the cell line, it is conceivable to extend the exposure time, within the time of the cycle,

TABLE 20.2
Design of *In Vitro* Experiments to Mimic Corresponding Effects of Chemicals *In Vivo* at Cellular Level

Mechanism of *In Vivo* Effect	Design of Corresponding Repeated Chronic Exposure Studies *In Vitro*	Objective
A. Slow accumulation of chemical in target tissues until acute toxic concentrations are reached	Cultures exposed to *low but increasing doses* of chemical for at least two to three passage levels	To determine whether chemical causes toxicity at concentrations below threshold or accumulation is necessary for toxic response
B. Slow increase in blood concentrations that subsequently distribute to target tissues	Prolonged repeated intermittent exposure experiments; proliferating cells are exposed to continuous *low* doses of chemical for three or more passages	To compare corresponding toxic responses between accumulation at low concentrations and steady state concentrations (as in A.)
C. Chemical does not accumulate but causes toxicity through repeated insults until threshold injury results	Repeated intermittent dosage assays: short repeated exposure to test chemical (up to 24 hr) followed each time by substitution with fresh medium; incubations repeated in each of three successive passages	To determine extent of insult occurring at cellular level in absence of agent, after repeated insult by allowing sufficient time for uncoupling of non-covalently bound substance

to determine the response to a chemical. In this way, exposure to the chemical through at least one cell cycle differentiates between acute and subacute or subacute and chronic. This sequence for exposure also reveals any tolerance developed against a chemical.

The effects of chemicals *in vivo* at the cellular level also result from repeated exposure and follow familiar patterns, including (1) the slow accumulation of the chemical in target tissues until acute toxic concentrations are reached, (2) the slow accumulation of the chemical in blood which subsequently distributes to target tissues, and (3) the chemical does not accumulate but causes toxicity through repeated insults until threshold injury results. To address these mechanisms, repeated and chronic exposure assays are performed, but are carefully outlined so as to mimic the *in vivo* situation. Examples of such experiments are outlined in Table 20.2.

20.2.5 RESULTS AND MECHANISTIC IMPLICATIONS

Basal cytotoxicity from a chemical, therefore, echoes the status of cell membrane integrity, mitochondrial activity, or protein synthesis, since these are examples of basic metabolic functions common to all cells. The level of activity, however, may vary from one organ to another, depending on its metabolic rate and contribution to homeostatic mechanisms, but the processes do not differ qualitatively among organs. In many cases, target organ toxicity in mammals reflects basal cytotoxicity of chemicals distributed to the corresponding organ. If this hypothesis is true, tests using continuous cell lines (e.g., 3T3 mouse embryonic fibroblasts) cover a large percentage of toxic effects and reduce the need to introduce many laborious systems with specific cells.

Organ-specific primary and continuous cultures are incorporated into tests for systemic toxicity. These cultures are capable of demonstrating basal cytotoxicity, similar to the effects shown by undifferentiated cells, without the need to measure specific functional characteristics. Organ-specific cells also express genuine site-specific cellular toxicity. A chemical capable of altering a particular function that is unique to an organ may challenge the basal or organ-specific origin of the mechanism of action. Comparison of the concentrations of chemicals in culture necessary to produce cytotoxic effects in cell lines of different origin may elucidate the underlying mechanisms of the toxic response.

The use of organ-specific cells to test for organizational toxicity is also suitable as a screening system, since these cells provide information on basal and specific cytotoxicity. For example, primary cultures of hepatocytes are used for the determination of basal cytotoxicity and organ-specific toxicity, based on the extensive metabolic capacities of liver cells to screen most circulating xenobiotics. Of concern with the use of hepatocytes in particular and many organ-specific primary cultures in general is that they are not routinely maintained in continuous culture and must be established for each experiment. In addition, primary cultures are more expensive than continuous cell lines and raise legitimate ethical questions when animals are needed as the sources of cells for repeated experiments. Thus, a long-term objective of efficient cytotoxicity testing models is the determination of the role of continuous cultures originally derived from specific organs and their relationship to similar cells in primary culture. Further exploration of this objective will elucidate how continuous cell lines reflect the *in vivo* organ's response to a chemical insult.

20.3 EXTRAPOLATION TO HUMAN TOXICITY

Twenty years ago, the fledgling perception that a standardized battery of *in vitro* tests could be developed for tier testing of acute local and systemic toxicity was speculative and ambitious. Progress in sensitivity, reliability, and technical feasibility of testing protocols supported by current validation studies is a testament to the realization of those original objectives.

20.3.1 STANDARDIZED TEST BATTERIES

Certain methods assigned to the battery are used for systematic mapping of the different effects attributed to basal cytotoxicity. The mechanisms underlying basal cytotoxic phenomena explain how these effects are influenced by the physicochemical properties of the molecules and how the knowledge is useful for the interpretation of results in cellular testing. Alternative cell-based models are applied on actions of chemicals known to be cytotoxic to animals or humans by comparing the *in vitro* and *in vivo* concentrations of the same chemicals.

Figure 20.2. summarizes the organization and applications of *in vitro* cytotoxicity testing and assigns positions for basal cytotoxicity, organ-specific cytotoxicity, and

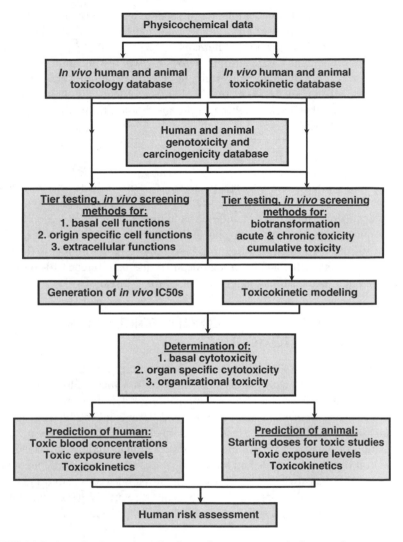

FIGURE 20.2 Organization and applications of *in vitro* cytotoxicology testing. (From Barile, Frank A., *Clinical Toxicology: Principles and Mechanisms,* CRC Press, 2004.)

organizational toxicity within the general scheme. According to the diagram, carcinogenicity and mutagenicity data contribute to the cytotoxicology evaluation but this information can be circumvented. Overall, the objective of this information is to arrive at human toxic blood concentrations to assist in the prediction of human toxicity and evaluation of risk assessment. Note also that *in vitro* toxicokinetic studies contribute to the establishment of the model. Together with *in vitro* concentrations derived from the tests, a model for cytotoxicity is formulated.

A battery of selected protocols functions as a primary screening tool before the implementation of routine animal testing. Subsequent animal tests are then performed to confirm the results from tier testing, as well as to detect outliers that elude the screening protocols. Upon completion of specific animal tests to confirm the *in vitro* data, a summary report of all available information is used as a basis for the screening and/or prediction of risk to humans.

With the refinement of general *in vitro–in vivo* testing and the progress afforded such studies with time, certain tests will be employed with greater confidence as it becomes apparent that the battery reflects the acute toxicity of certain groups of substances, resulting in the inevitable elimination of particular aspects of animal testing.

20.3.2 HUMAN RISK ASSESSMENT

Together with genotoxic and mutagenic tests, *in vitro* cellular test batteries are applied as biological markers of chemically induced risks, whether synthetic or naturally occurring. Such risks include biomonitoring of food and food additives and water and air analysis for environmental toxicants. If similar batteries of protocols are validated for their abilities to screen for or predict human toxic effects, these measurements will establish a significant frame of reference for monitoring of environmental, occupational, and commercial toxic threats.

20.4 USE OF ALTERNATIVE METHODS FOR HUMAN AND ANIMAL TOXICOLOGY

It is possible to apply the 3R principles to animal toxicity testing of existing chemicals in the environment with risk assessment based on the results of *in vitro* tier testing. This goal is achievable with a combination of data from reliable chemical and analytical measurements of human blood and tissue concentrations of the same chemicals, such as with volunteer testing of dermal and ocular irritancy or through the tabulation of databases with clinically relevant human toxicity information.

Based on observable trends and the practical experiences of many laboratories and regulatory agencies with *in vitro* cytotoxicity tests, it is conceivable and realistic that toxicity testing in the foreseeable future will be regularly performed with cell culture models. Furthermore, such testing will be more predictive of human toxicity than the current animal tests.

With the realization of this objective comes the benefit that many new types of *in vitro* toxicology and toxicokinetic tests will be available and many of the current methods will be validated in refined programs. These programs are conducted as refined multifactorial model systems of large *in vitro* databases to account for human toxicity, including computerized physiologically based kinetic modeling. Simultaneously, the gradual acceptance of new *in vitro* methods should not only depend on formal validation programs, but may be a consequence of the parallel experiences of various academic and industrial laboratories using both *in vitro* and *in vivo* methods. Finally, these achievements are not based only on political or societal attempts to reduce, refine, or replace the use of animals in research, but complement a very efficient and reliable system for protecting the public interest.

SUGGESTED READINGS

Barile, F.A., Mechanisms of cytotxicology, in *Introduction to In Vitro Cytotoxicology: Mechanisms and Methods*, CRC Press, Boca Raton, FL, 1994, p. 27.

Ekwall, B., Basal cytotoxicity data (BC data) in human risk assessment, in *Proceedings of Workshop on Risk Assessment and Risk Management of Toxic Chemicals*, National Institute for Environmental Studies, Ibaraki, Japan, 1992, p. 137.

Ekwall, B., The basal cytotoxicity concept, in *Alternative Methods in Toxicology and the Life Sciences*, vol. 11, Goldberg, A. and van Zupten, L.F.M., Eds., Mary Ann Liebert, New York, 1995, p. 721.

Garthoff, B., Alternatives to animal experimentation: the regulatory background, *Toxicol. Appl. Pharmacol.*, 207, 388, 2005.

Grindon, C. and Combes, R., Introduction to the EU REACH legislation, *Altern. Lab. Anim.*, 34, 5. 2006.

Gupta, K., Rispin, A., Stitzel, K., Coecke, S., and Harbell, J., Ensuring quality of *in vitro* alternative test methods: issues and answers, *Regul. Toxicol. Pharmacol.*, 43, 219, 2005.

Koeter, H.B., Dialogue and collaboration: a personal view on laboratory animal welfare developments in general, and on ECVAM's first decade in particular, *Altern. Lab. Anim.*, 30, 207, 2002.

Schechtman, L.M., Implementation of the 3Rs (refinement, reduction, and replacement): validation and regulatory acceptance considerations for alternative toxicological test methods, *ILAR J.,* 43, S85, 2002.

Schechtman, L.M. and Stokes, W.S., ECVAM-ICCVAM: prospects for future collaboration, *Altern. Lab. Anim.*, 30, 227, 2002.

Straughan, D., Progress in applying the three Rs to the potency testing of botulinum toxin type A, *Altern. Lab. Anim.*, 34, 305, 2006.

REVIEW ARTICLES

Combes, R., Gaunt, I., and Balls, M.A., Scientific and animal welfare assessment of the OECD Health Effects Test Guidelines for the safety testing of chemicals under the European Union REACH system, *Altern. Lab. Anim.*, 34, 77, 2006.

Edler, L. and Ittrich, C., Biostatistical methods for the validation of alternative methods for *in vitro* toxicity testing, *Altern. Lab. Anim.*, 31, 5, 2003.

Gruber, F.P. and Hartung, T., Alternatives to animal experimentation in basic research, ALTEX, 21, 3, 2004.

Huggins, J., Alternatives to animal testing: research, trends, validation, regulatory acceptance, *ALTEX*, 20, 3, 2003.

Kirkland, D.J., Testing strategies in mutagenicity and genetic toxicology: an appraisal of the guidelines of the European Scientific Committee for Cosmetics and Non-Food Products for the evaluation of hair dyes, *Mutat. Res.*, 588, 88, 2005.

Louekari, K., Status and prospects of *in vitro* tests in risk assessment, *Altern. Lab. Anim.*, 32, 431, 2004.

Meyer, O., Testing and assessment strategies, including alternative and new approaches, *Toxicol. Lett.*, 140, 21, 2003.

Rispin, A., Stitzel, K., Harbell, J. and Klausner, M., Ensuring quality of *in vitro* alternative test methods: current practice, *Regul. Toxicol. Pharmacol.*, 45, 97, 2006.

Stokes, W.S., Schechtman, L.M., and Hill, R.N., The Interagency Coordinating Committee on the Validation of Alternative Methods (ICCVAM): a review of the ICCVAM test method evaluation process and current international collaborations with the European Centre for the Validation of Alternative Methods (ECVAM), *Altern. Lab. Anim.*, 30, 23, 2002.

Index

A

Absolute risk, teratogenicity, 117

Absorption (toxicokinetics), 28–34
   animal tests, descriptive, 67
   Henderson-Hasselbach equation, 30–32
   intestinal tract, lipid/water partition
      coefficients and, 37
   in vitro ADME studies, 206–207
   ionic and non-ionic principles, 28–30
   in nasal and respiratory mucosa, 32–33
   transport of molecules, 33–34

Absorption, distribution, metabolism, and
      elimination (ADME) studies, 28,
      206–207

Academic research, applications of toxicology, 5

Acceptable daily intake (ADI), developmental
      toxicity, 117

Accumulation, chronic toxicity, 90–91

Acid/base characteristics
   Henderson-Hasselbach equation, 30–32
   ionic and non-ionic principles, 29–30
   plasma protein binding, 37

Actinic keratoses, 106

Active transport, 33

Activity, plasma protein binding and, 37

Acute effects
   delayed effects versus, 14
   reversibility of, 14

Acute exposure
   exposure duration and frequency, 18–19
   subchronic, defined, 90

Acute lethality determination, LD_{50} tests, 83–86

Acute toxic class method (ATCM), range-finding
      tests, 77–78

Acute toxicity
   alternative methods for measurement,
      284–285
   in vitro tests, 192–200; *see also* Cell culture
      methods for acute toxicology testing
      dermal, 198–200
      ocular, 193–198
   in vivo tests
      dermal, 101–107
      ocular, 107–108, 109

Acute toxicology tests/testing, *in vivo* (LD_{50} tests),
      73–86
   applications of LD_{50} studies, 86
   classical LD_{50}, 78–80
      dermal, 79–80
      inhalational, 80
      oral, 78, 79
   LD_{50} and, 73–74
   objectives, 73
   organization of studies, 75
   other considerations with LD_{50} tests, 80–86
      biological variation, 83
      determination of acute lethality, 83–86
      duration, 82
      general appearance of animals, 82
      routes of administration, 81–82
      specimen collection and gross pathology,
         82–83
   range finding tests, 75–78
      acute toxic class method (ATCM), 77–78
      fixed-dose procedure (FDP), 77
      up-and-down procedure (UDP), 76–77

Additive interactions, 20

Additives
   classification of toxic agents, 7
   and reproductive toxicity, 111

Ad hoc analysis, 263

Adipose tissue accumulation
   chronic toxicity, 90–91
   lipid solubility and, 37

ADME studies, *see* Absorption, distribution,
      metabolism, and elimination
      (ADME) studies

Administration/exposure routes, 15–18
   acute toxicology tests/testing, *in vivo* (LD_{50}
      tests), 73, 78–80
      dermal, 79–80
      inhalational, 80
      LD_{50} tests, 81–82
      oral, 78, 79
   animal tests, descriptive, 60, 67
   carcinogenicity testing *in vivo*, 136, 139
   dermal and parenteral, 17–18, 79–80
   exposure assessment, 47
   inhalation, 17, 80
   intranasal, 17